# Global Childhoods

## Globalization, Development and Young People

Edited by Stuart Aitken, Ragnhild Lund and Anne Trine Kjørholt

Routledge
Taylor & Francis Group

LONDON AND NEW YORK

First published 2008 by Routledge
2 Park Square, Milton Park, Abingdon, Oxon, OX14 4RN

Simultaneously published in the USA and Canada
by Routledge
270 Madison Ave, New York NY 10016

*Routledge is an imprint of the Taylor & Francis Group, an informa business*

Transferred to Digital Printing 2009

© 2008 Stuart Aitken

Typeset in Times Roman by Techset Composition, Salisbury, UK

*British Library Cataloguing in Publication Data*
A catalogue record for this book is available from the British Library

*Library of Congress Cataloging in Publication Data*

ISBN 10: 0-415-41145-9 (hbk)
ISBN 10: 0-415-49488-5 (pbk)

ISBN 13: 978-0-415-41145-5 (hbk)
ISBN 13: 978-0-415-49488-5 (pbk)

# Contents

# Foreword

Childhood studies is, by now, a well established field of multidisciplinary enquiry with questions about children's everyday lives, children's voices and the diversity of conceptualizations of childhood at its core. Given the earlier dominance of discourses of childhood that saw children as simply smaller, less capable versions of adults and as people in need of care and control, and of childhood as the unproblematic period of early human development, that childhood studies is now flourishing is testimony to the tenacity of researchers in the field who doggedly pursued their alternative agenda. However, this is no time to be self-congratulatory for while childhood studies might have come of age, as this book amply demonstrates, it is not yet fully matured. There is still much work to be done.

In this volume we begin to see some of the ways in which the project of childhood studies can be further progressed through the questions it raises about children and (their) place. It asks not just about children's participation in local communities but on the world's stage; it asks about the existence of a global childhood and whether this is a singular or multi-faceted childhood; it asks about the extent to which children can ever be shielded from the larger structural forces that shape the world's economies and, necessarily, in doing so, what impact these have upon the lives of children. These are 'big' questions that deserve more attention than hitherto has been given to them, consistently, from within childhood studies.

Of particular importance is the new spin placed on the question of whether 'childhood' is universal or not. No longer is this simply a debate with traditional developmental psychology but, as we see here, it is a question of whether global market forces now mean that children, as consumers of media and other technologies, potentially share in a common culture. Of significance too is the global interconnectedness of children's everyday lives. Children's cultural identities may no longer be as 'localized' as they were once thought to be since migration and children's work experiences can enable different kinds of self-transformation that are not bound to specific locales.

Discourses of childhood and children's participation also come in for reconsideration through this volume's attention to the different ways in which 'ideas' about what children are or should be become used as vehicles for other national, political agendas—indeed, they may not even serve children's interests at all! Children's 'presence' as participants may not be required for it is the idea of 'the child'—rather than real children—which carries the greatest symbolic weight in such arenas. On the other hand, as this volume also reminds us, 'real' children can suffer from the weight of that symbolism when, in the world's global markets, the everyday contribution that working children make continues to be under-recognised and under-valued.

For childhood studies, therefore, this volume marks another step in highlighting the contribution that it has to make not only to understanding the lives of children but to also understanding the ways in which the study of children and childhood has a great deal to offer to the development of social theory more generally.

*Allison James*

# Why Children? Why Now?

STUART C. AITKEN, RAGNHILD LUND AND ANNE TRINE
KJØRHOLT

*Stuart C. Aitken, Department of Geography, San Diego State University, San Diego,
CA 92128, USA*
*Department of Geography and the Norwegian Centre for Child Research, Norwegian
University of Science and Technology, 7491 Trondheim, Norway*
*Ragnhild Lund, Department of Geography, Norwegian University of Science and
Technology, 7491 Trondheim, Norway*
*Anne Trine Korholt, The Norwegian Centre for Child Research, Norwegian University of
Science and Technology, 7491 Trondheim, Norway*

We are inspired to write about children in our contemporary globalized moment for a variety of compelling reasons. Although evoking the notions of globalization and children's well-being may run the risk of pivoting discussion on terms that are over-used politically from both the left and right, they nonetheless suggests as a beginning a broader context for a discussion about how young lives are elaborated. As a process, globalization may have been around for a while, and yet today the term suggests rapid change and connectedness at a number of important levels. It seems that local economic, political and cultural contexts respond much more quickly to processes that arise elsewhere, market adjustments are close to instantaneous and, with regard to the ideas contained in this volume, in a world connected by flexible capital, mobile labor and transnational families, young people provide an important fulcrum of, and impetus for, change.

Adjustments in terms of the resiliencies, empowerments, oppressions, resistances and manipulations of young people in the face of social, cultural and economic structural transformation provide focus for this volume. Our writing is not about rescuing children, taking them out of labor and putting them in classrooms, or trying to raise the standard of well-being in those parts of the world that fall behind the so-called development curve, nor is it about creating a better world for the next generation; indeed, part of our dispute with these kinds of rhetorical arguments is that they are couched in problematic progressive and developmental terms that foist adult agendas on young people. Rather, the authors in this volume recognize the myriad contributions of young people to global processes and the many ways that those contributions are hidden, subverted, or contrived through adultist machinations at the global and local level. The volume is about re-setting the ways we understand these contributions from post-development perspectives, which challenge not only the conventional wisdom on how places and economies change but also how young people develop through and within these changes and how they too are agents of change.

The work that follows is based on discussions that began in San Diego, California in Fall 1998 and continued in Trondheim, Norway in Spring 2005. The focus of the San Diego discussions was on our understanding of the ways young people inter-relate with their social and spatial contexts (cf. *Environment & Planning A*, 2000; Holloway and Valentine, 2000; Aitken, 2001; Ruddick, 2003). At that time, the influences of globalization suggested an enduring and important context for the lives of young people and the extent of a myriad of global childhoods, and so these themes were taken up in the workshop that followed in Trondheim. The current volume collects together individual contributions that built upon the global childhoods discussions. Collectively, these papers turn on its head the idea that young people are merely the recipients of global progress. Rather, new ways of understanding reproduction and geographies of economic development suggest that the lives of young people are increasingly important for understanding larger notions of change. Furthermore, it is clear that children are actors and competent arbiters of change even in situations of exploitation. These new ways of understanding children in the world are slow to penetrate the seeming wisdom of adult solutions. It is clear that despite young people's dominance as a demographic category in the majority world—including their influence globally as a market niche and their importance as a focus of care and responsibility—they, and their voices, are still largely missing from larger academic debates on globalization.[1] There is a tired inevitability to the progressive rhetoric of academics and policy-makers that, we argue, requires a re-setting, a complete turnaround, if you will, of the ways we come to know children and development in the world. The stark and oppressive outcomes of neo-liberal agendas and global corporate capitalism—such as the commodification of lifestyles, a global sex trade, new diasporas and wars that increasingly involve child soldiers—on the world's young people are not inevitable.

There are a number of recent texts that explore the global contexts of children and young people (Katz, 2004; Ansel, 2005; Goddard *et al.*, 2005), but few address directly and critically the complex agendas that propel notions of both child development and economic development. The ways we come to know the so-called development of children and economic development are now significantly changed by perspectives that are collectively known as post-development studies. 'Why children? Why now?' is not just about noticing young people in the full ambiguity and complexity of global and local context (to use Buckingham's words in this volume), it is also about rethinking the global spaces of young people as grounded and lively, full of promise and surprise and, thus, open to the political.

The volume is organized into three parts. The first part, comprising essays by Gagen, Kjørholt and Buckingham, focuses on the ways that childhood is constructed in the north and then exported globally. The key issues are play, education, authenticity and representation. The second part, comprising essays by Bosco, Tatek Abebe, Punch and

Aitken, focuses on the necessities of the global south and young people's creation of livelihoods. The key issues are child labor, and geographies of care and responsibility. The third part, comprising essays by Nieuwenhuys, Lund and Skelton but embracing much of the intent of the other essays in the volume, furthers the possibility of children's participation in, and construction of, a very different form of globalized (and yet intrinsically and materially local) development. With this structure for the essays set up as a pedagogic apparatus, we spend the rest of this introductory essay—*Why Children? Why Now?*—with the larger concepts that we believe this volume is radically rethinking, using the essays of the contributors as foils upon which we hope the current inevitability of global neoliberal progress loses its power.

## Why Rethink Development?

We argue that a rethinking of economic (and cultural) development and its relations to global spaces is concomitantly a rethinking of progress through this thing that is called childhood. Our premise is that the two literatures of post-development studies focusing, on the one hand, on getting beyond traditional notions of economic development and, on the other, challenging child development as a series of stages need to come together if we are to confront the excesses of neoliberal capitalism and the disempowerment of children. This volume represents a halting and yet passionate attempt to bring together the notions of the development of children and economic development in a critique that finds coherency in its attack against a set of academic and policy enframements that, we argue, limit possible futures. It is a radical attempt to re-imagine young people in a less bounded world. The arguments are rehearsed in contemporary social, critical and post-development theory, but this is the first time they have been brought together in one volume. Briefly, the critique is against the notion of stages of development for either children or nations, and the problematic political results when the two are equated together.

We argue that progress and development are almost always the products of teleological closure. Interest in evolution, continuum and/or movement cannot be broken down into discrete stages. And yet, as Gagen and Nieuwenhuys show, there is a long history to this kind of project and its global mapping falls on the shoulders of the academic community in problematic ways. We are concerned about opening up the political for young people, and we believe that change is only possible with movement away from static time-slices and spatial emplacements elaborated by academics and embraced by policy-makers.

For the most part, we argue, developmental theory essentializes children to become-the-same (as us) with limited possible futures. Similarly, so-called developing nations are cajoled and coerced to become–the-same (as us); and this is a fantasy that remains unrealized from their peripheral and exploited positions in the world. These positions construct children's and nations' contexts as no different from ours, and it denies (or at least hides to a degree) our need for their relatively cheap labor and resources. In so doing, we deny them their own trajectories, their own histories, and futures that are different, and perhaps better, than ours. We deny them the possibility of becoming-other: they are merely at an earlier stage of the fantasy that we create, which, in actuality and solely, supports our rich lifestyles. What we just offered, of course, is yet another problematic framing and the papers by Lund, Aitken, Kjørholt and Bosco, in particular, are concerned about foundational and problematically dichotomized concepts such as us/them, local/global, global north/global south. As Lund points out, when the complexities of the local are looked at closely it is difficult to find a global south.

And so we are wary of framings that are encapsulated within traditional representations. In this sense, representation is reification, a static fixing of things. In the essays that follow—and particularly those of Buckingham, Skelton, Aitken and Bosco—there is a critique of this kind of boundedness, and an attempt to dislocate young people from spatial/historical framings that produce static representations. Instead, we introduce another set of ideas that focus on fluidity, movement, relationality, multiplicity and liveliness. It is an attempt to understand the worlds of young people as neither composed of monastic individuals, nor closed off in categories or stages. Rather, we look towards openness and incoherency as hallmarks of the radically political and the actively experimental.

Before embracing openness, it is important to understand the roots of our current framings. Child development and economic/cultural development are projects with long, complicated and inter-connected histories. In this volume, Gagen looks at development historically, taking us back to the beginnings of modernity with the nineteenth century's elaborated hegemony of empire, to a time when colonialism and imperialism expanded the control of so-called metropolitan heartlands to a colonial periphery. This expansion saw space as something to be controlled and history as a developmental given. Peripheral colonial spheres were to be civilized with the imposition of governmental, legal, economic and educational frames from the metropolitan heart. Gagen elaborates the way that the US, in the late nineteenth century, came into the realm of imperial exploitation. The link to notions of child development is clear here. The problematic spatial framing that Gagen elaborates focuses on domestic changes propelled by psychology's suggestion of 'normal' development. Of some considerable importance to the arguments that this volume makes is Gagen's suggestion that the infantilization of so-called primitive cultures is, in actuality, a discourse derived from the normative liberal developmentalism of psychology, which finds its beginning in a western bourgeois understanding of children as inherently primitive. Progressive attitudes to child development were not only instilled at home but they were also exported from the US to its imperial protectorates through the establishment of schools, physical education programs and playgrounds abroad. Gagen argues that these two strands of development—one internal and one external—are all part of the same imperial project. Like nascent colonies of the US, children had to move along a normalized path from underdevelopment to development. In the nineteenth century, the newly introduced science of child development paved this path. Importantly for what this volume is trying to do, Gagen points out that the notion of 'underdevelopment' was created on a geographical and temporal plane, co-constituting foreign protectorates and children to a universal state of pre-modern, primitive and in need of protection and paternalistic guidance.

These early notions of development are elaborated in the twentieth century by UCLA economist, Walter Rostow, who proclaimed in the 1960s a systematic and progressive view of national development from premodern, traditional societies through a series of transitional stages to what he called high consumerism. The Rostovian take-off model was criticized early on for an insular and locally embedded teleology that had little to say about global inequalities in access and resources and internal social injustice. Rostow's work nonetheless was celebrated as a tonic to Marx's progression to communism, and as an optimistic solution for so-called undeveloped and underdeveloped economies. As such, Rostow's work influenced a generation of economists, sociologists, demographers and geographers interested in development issues. Models such as the demographic transition theory similarly predicated development in terms a few causal variables and seemingly inevitable stages. Despite significant critiques (e.g., Teitelbaum, 1975), the demographic transition model continues to hold academic attention as a theory of world development (Dyson, 2001).

Kate Willis (2005) articulates a long chronology of approaches to understanding economic development, from modernist theory (e.g., Rostovian take-off theory) and neo-modernist/liberalist perspectives (including Wallerstein's 1980 influential world systems theory) to post-structural and post-development critiques that focus on grassroots activities and local level participation (e.g., Escobar, 1994), but also to the more recent post-modernist and post-colonial movements (Pieterse, 2001). In this volume, Lund tasks the appropriateness of alternative developmental models to issues that relate to child participation, which she calls an overstretched analytical and practical term.

There are important links between the varied ways of conceptualizing development that are quite troubling. Modernist perspectives dovetail with neoliberalism, for example, in that both suggest the inevitability of globalization and the hegemony of multi-national corporate agendas and consumerist ideals. Given this logic—which we decry as positing an inevitability that forecloses upon the political—then the pinnacle of development for both children and nations is a stage of high consumerism and, as such, Buckingham's essay provides an ironic elaboration of child/media relations with an understanding of global capital seeking to expand consumer niches globally towards the highest possible market gain. The context of the local and the global figures hugely in his concerns, and it highlights the broader concern of this volume on what constitutes the local as something more than defensible space and globalization as something inevitable and 'out there'. On the one hand, and in relation to Gagen's historical account of imperialism emanating from the US, Buckingham focuses on theories of contemporary cultural imperialism from Hollywood. The hegemonic rise of Disney as an icon of representations for children, argues Buckingham, is not a sufficient account, because it implies that the local is powerless to resist ideological domination. He favors a notion of a globalization that enables local cultures to flourish. The question is thus raised of children's local cultures in the face of a global media onslaught. And yet this too is a limited view, because access is not universal. Citing the current rise of Japanese anime, the anxiety that Buckingham raises relates to possibly diminished cultural continuity and intergenerational socialization with increasingly homogenized global media for children. Importantly, Buckingham demonstrates the unpredictable and contested relationships between the local and global, and the consequences for children's culture.

## Scale, Speed and Adjustment

It is important to note that discourses on development are tricky because they may be considered at different scales and rates of change, ranging from the individual, the local community, the on-track and the speeding, the regional, the tardy, the dismissed and the globally-connected. Willis (2005) points out that Keynesian approaches, for example, are structural state interventions to help regions and groups that are disadvantaged whereas dependency theories focus on global economic inequalities that are perpetrated on the periphery as a result of exploitation from the global north. Neo-liberal agendas are based on speedy private sector interventions that offset the seeming enervation of state interventions. The notions of liberal democracy embedded in these latter agendas, notes Willis, see the state as providing a regulatory framework that enhances the speed with which corporations and non-governmental agencies (NGOs) act. Unfortunately, as Lund shows, NGOs and CBOs (community-based organizations) very much represent a status quo and their transformative potential for children is limited. That said, she argues with Aitken and Punch that children arise as important agents of change in many contexts, including those that are potentially exploitative and abusive.

Buckingham notes that a large context of the so-called 'modernization' of nations and children is access to global media. The ties of young people to larger global representations are also a large part of the essays by Bosco and Aitken. Bosco is concerned about the ways that local contexts for children are rapidly 'unhooked' and appropriated by trans-local networks of aid and solidarity. Focusing on two internet-based NGOs, *Por Los Chicos* and *Red Solidaria*, he is concerned about the ways that they try to care for others, particularly those who are distant. The Argentinian debt crisis of the 1990s is particularly pertinent to his work. Bosco argues that diminished support for *Por Los Chicos* over the years suggests the problems of disembodied connections. Part of this is about representation and how this web site was set up, which revolved around problematic conceptions of the universal child and universal notions of care. These conceptions focus on a decontextualized child (wide-eyed and staring at the camera) who, seemingly, can be helped only by the viewer (cf. Ruddick, 2003). This is similar to Aitken's argument, which extends decontextualization (which he terms 'ghoulish effects') to changes in representations of child labor to aid civic boosterism in Tijuana. Bosco likens this to Escobar's (1994) 'objectification of the visual'.

Aitken argues that institutionalized and sanitized child labor in Tijuana may be seen as another example of the ways local authorities clean up and modernize their space through changed representations and, at the same time, young people are tamed as they learn to labor. Aitken suggests a program whereby Tijuana's modernization is couched in a move away from young people selling chicklet's gum in tourist sites to young people actively serving as checkout packers in supermarkets. The creation of smart, smiling young people as grocery packers in clean, modern supermarkets is a relatively recent gambit for a border city that wants to be seen as a major a player on the world's economic stage. Aitken uses a post-structural experimental analysis to suggest the deliberation of a modernized aesthetic; young people who are not only smart and clean, but also willing to serve a consumerist ideal.

Although Aitken's work is tied to issues of representation it also highlights regional adjustments to global economic transformations that reflect changing labor patterns of young people. The issues of local and regional economic transformation and labor are also highlighted with the contributions by Tatek Abebe and Punch. Tatek Abebe focuses on post-development contexts of child labor in rural Ethiopia. He argues that development strategies fueled by globalization are failing the region's young people. The main problem is an increasing focus on cash-based coffee growing over the last few decades at the expense of subsistence crops. Liberal free-trade policies do not distinguish high quality beans in his study area from cheap mixes from elsewhere. Structural adjustment programs exacerbate the problem and result in young people shouldering the bulk of local reproduction at the expense of their schooling. This context is not sufficiently complicated, Tatek Abebe argues, by an understanding of indigenous, Islamic and colonial factors. He suggests a material grounding in global space, outlining the seasonality-, age- and gender-differentiated work of young people, and pointing out important relations to global market forces.

Punch also points to global market forces using an empirical study of young migrants from Bolivia to argue for a renewed focus on the lack of consequence of education in their lives. She argues that rather than formal education, it is their movement to Argentina on a seasonal basis that enables young people to develop their global interconnections and to increase their status back home. As an empirical study, Punch's work is important because it focuses on the stories of the young people both at home and abroad, but nonetheless in a regional context. The problem of landlessness, or waiting for land to free up, forces many young people to seek better opportunities in Argentina. The proximity of the move (less than a day's bus journey) is important for keeping in touch with people at home. For example, first trips are often with friends or relatives, and parents can visit. In addition, the stories (and material goods) with which migrants return to Bolivia, entice other young people to migrate. Despite their storied elaborations, the life of migrants in Argentina is extremely hard, with long work hours and cramp living conditions. Punch notes the importance of material goods (particularly Argentine clothes) for these young migrants, which ties into a popular neo-liberal Latin American governmental rhetoric that consumerism is synonymous with freedom (seen also in Aitken's work). Migration for these young people is a coping strategy which facilitates their participation in a more consumer, and sometimes global, culture as well as enhancing their social and economic autonomy.

The relations of the local to the global in the last 40 years of development literature and policy are soundly critiqued by Arturo Escobar (1994, 2001), who points out the problematic ways that the local is linked to labor, tradition and vulnerability (including women, minorities, the poor and, of course, children). Escobar argues for the potential agency of local culture and local places in the context of neoliberal globalization. The work of Tatek Abebe and Punch (but also Buckingham, Aitken, Lund and Bosco) highlight some important contexts of local variation with the suggestion that the future of economic transformation and the wholesale betterment of the world through universal consumerism and material gain is questionable. And, in relation to this, the possible futures of young people within this context are not knowable and should not be foreclosed upon by suggesting that certain outcomes are inevitable. If we think of global space relationally then it is important to not reify it as out or up there, but rather as utterly grounded in material practices of young people in Bolivia, Ethiopia, Argentina and Mexico, and that those practices and connections, in the words of Doreen Massey (2004, p. 185) 'may go around the world'.

## Taming Young People and Their Spaces

Imagining space as a container of young people's activities or as a surface across which they travel is a denial of possibilities. It is an enframement that is not only limiting, it is the kind of framing that perpetuates the gross inequalities of contemporary neoliberal globalization. Similarly, turning space into time by focusing on progressive child development denies young people new ways of configuring their lives. This is a way of taming the challenge of creating new, liberatory stories. Early work by Erica Burman (1994) and Valerie Walkerdine (1988) challenged the ideas of progressive and staged child development that became conventional wisdom with the work of Piaget and Inhelder in the 1950s (cf. Piaget and Inhelder, 1956). Like the work of Rostow in economics and despite significant empirical evidence to the contrary, Piaget and Inhelder and their students upheld the notion that children went through a series of intellectual stages that corresponded to their age. These notions held sway (and still are powerfully embedded) in the work of many educators, academics and policy makers.

The discipline of child development in the nineteenth century as described by Gagen becomes hugely indebted to Piaget's work in the mid-twentieth century. To counter this focus on staged development, Valerie Walkerdine (1984) argued that young people develop in more idiosyncratic and nuanced ways. Her work focused on classroom behaviors and showed that a large part of the developmental studies that reified Piagetian structures emanated from the ways classrooms were designed and curriculum formulated. In short, the sense of a place (e.g., how a teacher decorates a classroom) and the construction of a space (e.g., the arrangement of desks) matters to how young people came to know the world. Eric Burman (1994) furthered Walkerdine's concerns, arguing for a post-structural elaboration of child development that removed the problematic spatial enframement of specific stages. What all the papers in this volume attempt to do in a variety of implicit and explicit ways is to look closely at how notions of child development are not only problematic, but are also linked to the way economic and national development is conceived.

The notion of development as a series of stages through which children progress to a predetermined, normal adulthood resonates with the notion of a nation's development in a global market economy. Like time, space becomes an expanse to cross, the promise of development, of new markets elsewhere, of the global south awaiting the arrival of global capital (from over here), awaiting an understanding of what it is to grow up. This smacks of imperialism and sounds a lot like the creation of lands to conquer and colonize. The creation of times of childhood (for normal development) and spaces for children (for free play and schooling) are similar kinds of enframements. And there are similarly hidden agendas. Skateboarding is outlawed on public streets while at the same time a market niche is created for costly skateboard parks. This smacks of imperialism and the conquering and colonization of the bodies and minds of young people. The inevitability of a progressive form of neoliberal global capitalism with its penchant for universal individual rights (rights to education and skateboarding but only within specified institutionalized spaces where corporations are equated as individuals and given the same rights) suggests a problematic move towards inequality. An inequality is created spatiality that disadvantages minority populations, such as less well off children who cannot afford market-priced education and skateboard parks.

What this comes down to is suggestive of Massey's (2004, p. 5) point that in striving to convince us that globalization is inevitable there is a problematic sleight of hand whereby geography is turned into history and space into time in the name of progress. Neoliberal globalization as a material practice is 'yet another in a long line of attempts to tame the spatial' and, as such, it evades 'the full challenge of space' (Massey, 2004, p. 99). Ethiopia, Mexico and Bolivia are not geographically disadvantaged because they are bequeathed the hope of development. They are bequeathed a particular history (our history) that negates their particular local geography. This, of course, has social and political effects. It says that so-called developing countries are not really different from us, and it denies the need of global capital to maintain inequality. Similarly, young people are formed as fodder for a larger global enterprise within which they have few choices. As Lund argues through a discussion of the Sri Lankan civil war and diaspora, children's lives and situations are articulations of the global in terms of a *time-space distanciation*. She points out that young people's displacement to labor elsewhere and their conscription/abduction into rebel armies represent examples of globalized 'material practices' that provide few choices for young people.

### The Bind of Universal Children's Rights and Global Discourses

The issue of more choices for young people is intriguing and hotly debated within the context of universal children's rights. The argument that we critique here is that universal rights emanates from a progressive neoliberal individualism that assumes identities are always already constituted and fixed. Claims for equality come from a pre-given understanding of what a child is; what we want to do here is replaced this rhetoric with a politics that takes the constitution of young people's identities and the relations through which they are constructed as a central concern. In terms of global politics, these relations are understood as embedded material spatial practices. Rather than accept the pre-given identities of young people, this politics elaborates the relational constructedness of things. It is hugely wary of claims to right and truth based on the seeming authenticity of unchanged political identities. Similarly, space too is a product of relations, and does not exist prior to identities and their relations. There is, by this way of thinking, an important connection between the co-creation of space and young people.

Spatial identities (nation, border, place) are tied intimately to political identities (adult, adolescent, student, child). The issue of contemporary spaces is forwarded by Kjørholt's focus on the impossibility of authenticity in children's voices and the undaunted search for it by some national and global institutions. The project of neoliberal globalization finds form for her in the UN Convention on the Rights of the Child (UNCRC), which has generated heated discussion and some problematic policy making at the national level over the last 15 years. This is an issue that Lund takes on from the perspective of children's participation. The UNCRC's penchant for universal rights places it squarely in the realm of a neoliberal agenda that fixes individual categories of existence/identity. Kjørholt takes this to task from the perspective of the place of children in society as social actors in everyday life. Importantly, she argues that children reproduce life on par with adults and, as such, they are co-creators of their childhoods. The question remains as to how much license they are given in their part of the creation. And, moreover, how much this participation is hidden by liberal developmentalism. It seems also that right discourses contribute to this invisibility by universalizing the contribution of young people. Kjørholt argues that new liberatory discourses may be emerging from the UNCRC that open up space for new forms of child participation in social and political reproduction at both the national and international scale. She looks at two programs, one in Norway and one in Denmark, which suggest changing participation with a focus on children as social actors actively reproducing culture and national identities. By looking at individual stories of children, Kjørholt demonstrates the tensions between the enframing of global discourses (such as the International Labor Organization's mandate to stop child labor as discussed by Aitken and Nieuwenhuys) and the realities of children's lives. The global rhetoric glosses over the places of young people and suggests a problematically straight forward progress to development. As such, many children (and particularly the child activists touted by the UNCRC), are what Kjørholt calls 'symbolic participants' in a larger neoliberal agenda.

The Norwegian project discussed by Kjørholt suggests a shift in focus from the 'developing' child to the 'competent' child; from 'pedagogy' to 'culture'. The notion, however, is that a culture made by children (children's own culture or play culture) is rapidly disappearing. This larger discourse may not be so far removed from Gagen's concerns. The question of how this relates neoliberal agendas to the notion of participatory democracies wherein yet another group is given so-called freedom is interesting. Freedom, as another part of the neoliberal agenda (Harvey, 2005), is highlighted in the free activities of Kjørholt's Danish case and, in both the Norwegian and the Danish case, development is constructed as something that is politically and ideologically neutral. Once again, as Kjørholt points out, the issue relates to adult needs for a nostalgic reconstruction of their own childhoods.

## Unfettered Global Childhoods

Looking at national and international discourses from a different perspective, Nieuwenhuys equates the hidden reproductive work of young people with what she calls 'the global womb'. Her work is also about the ways representations produce static stages of development. She argues that the rhetoric of the child labor abolitionists are mythic and highlight ever receding notions of a better life. She argues that with neo-liberalism, any advantages accruing to the collapse of colonialism were lost to children in the majority world with the rhetoric of 'the best interests of the child', where 'interests' were defined as self-interest to fully participate in a market approach to development (see also Punch and Lund). Nieuwenhuys argues that, by so doing, symbolic violence is practiced against the notion of the non-consuming child and their lifeworlds, a perspective that fully endorses Buckingham's anxieties.

Nieuwenhuys' rethinking of child labor begins with the premise that the current need to eliminate the labor of young people is because of something that 'development' made problematic with regard to the production of life. Historically, people lost tenure of land and so children were no longer needed as repositories of family wealth and the new industrial wages were not sufficient to support them, so young people emerged as a 'social problem'. This problem was solved in part by removing children (and women) from the work-place and creating a so-called family wage for men. Reproduction was no longer at the expense of the industrialist. Like Gagen's concern with the emergence of the science of child development, Nieuwenhuys' story is global in scope, highlighting changes for young people's labor in the metropolitan core and in the colonial periphery, where great pains were taken to maintain the peasant family as the locus of a self-sufficient reproduction, which was not at the expense of empire. Nieuwenhuys provides examples from around the colonial globe of the ways children's work was redefined into an ethic that justified draconian measures of control (and continued cheap child labor). What this evidence suggests is that child labor elimination in heartland Europe and the US went hand-in-hand with new forms of child exploitation unfolding in the colonial periphery. In this sense, childhood was circumscribed territorially to the heartland and defined as 'other' (not really children) in the periphery. And so, as Aitken points out, concern with changes in child labor is also a concern that young people's identities are not conceived and represented as service to a larger corporate/civic ideal. Considering child workers in Scotland and Mexico over a span of 40 years, he argues that the 'ghoulish effects' of global discourse such as those of development and modernization contrive an aesthetic that plays out in the lives of young people. Young people's bodies and performative actions are bound to perceptions and representations as well as to the peculiarities of time and space.

Nieuwenhuys goes on to argue that child labor (and calls for its elimination) as a concept in the global south emerged in the 1990s as another form of labor control. As with other notions of children's practices, global child labor is overwhelmingly interpreted as a repetition of northern history and development, and it is dealt with solely within national borders. Collective memory about the role of child labor eradication in the making of the welfare state was sufficiently alive to align political response behind the common child saving agenda articulated by Kjørholt, but no welfare state was possible in a post 1990s neoliberal world. Nieuwenhuys argues that the lack of a national solution to this problem precipitated the notion of the global child. And it was seen as inevitable and necessary that children would suffer under national economic structural adjustments. Lund makes a clear case for this with her example of the Mahaweli project in Sri Lanka.

Given the laws of the neo-liberal market, Nieuwenhuys argues, consensus about the state's role in child labor converged around a set of practices that ritualistically celebrated children's rights, disengaged with social justice and, finally, put in place by social and economic programs that reformed the children themselves under close surveillance of northern donors. State control was diminished in favor of families, communities and the global society. The global child who is rendered for potential donors by the institutions such as the ILO and the UNCRC typically appears younger, is autonomously determined (like an adult), and wants only the restoration of lost adulthood.

Nieuwenhuys demonstrates how processes of a global rights approach, dis-embedded from a wider context, help justify states' disengagement from social and cultural reproduction in the direct interest of southern elites. This is the heart, she argues, of disciplining the global womb. And it is a heart which is populated by a lifeworld we know little about because much of the international rhetoric and research is focused upon what is appropriate and what is inappropriate work for children, and which dismisses their work to support families and economies through day-to-day activities as inconsequential.

In a similar vein, Skelton argues that we need to critically assess the UNCRC's focus on participation with particular concern for one of its sponsors, UNICEF, and its practice through the 2003 State of the World's Children report. Like Aitken and Nieuwenhuys, she is concerned with the ways that discourses from the north bleed inappropriately into young hearts of the global south. Skelton argues that the UNCRCs first set of emphases on protection and provision has of late been superseded by notions of child participation.

A large problem with these discourses of participation is that they are never used uncritically or unfavorably, and there is a presumption that child participation is always of value. Using critical discourse analysis, Skelton argues that the 2003 UNICEF report to highlight problematic focuses on child naivety, and their forward looking hopes and dreams. She shows how this rhetoric is problematically decontextualized and dismisses other avenues of participation, as noted also by Nieuwenhuys and Lund. The report is seen as a 'deepening of democracy', which is more inclusive and responsive but is also linked to a progressive form of development. Skelton then shows how the report shapes democratic citizenship within a tension of children 'becoming' adults (which is the discourse of the report) and 'beings' in and of themselves. This ties with the larger concerns of Aitken and Nieuwenhuys that young people and nations (their labor and their responsibility and care) are enabled to 'become other' in an future that is politically open rather than constrained to 'become-the-same' in a future that is prescribed by global neoliberal economics. The argument made here is that a large part of the globalized discourses from UNCRC and the ILO are clearly an enframing of children and an attempt to tame them.

Lund takes this a little further, positing problematically shifting discourses that certainly do enframe but whose effects are nonetheless difficult to pin down at the local level. She points out that in poor contexts, the participatory discourses of the UN and national governments are not only inadequate but perhaps also irrelevant. She uses case studies from Africa and Asia to suggest that although it is easy to document the ways participation is locally embedded, there are external structural forces related to globalization and geo-politics that are hugely relevant and much more difficult to pin down. To grasp this kind of complexity, she argues, we have to draw on more critical perspectives on the role of agency in recent developmental theory as well as insights from post-colonial scholars. Importantly, she points out that capabilities are not about what children can 'choose' but what they are able to achieve.

## Dissolving the Frameworks

In sum, our clarion call is for a realization and re-imagining of the full challenge of young people. A central part of the questioning in this revolves around the seeming inevitability of global development (and, in particular, its neoliberal construction) and children's development (and, in particular, their normative cognitive development). In this formulation, both globalization and child development are problematically enframed by force emanating always from 'elsewhere'. The normative stance in the essays that follow is that the frameworks of progress and development, for both young people and nations, must be replaced with something more fluid and politically open.

And so, as a basis for the rethinking that is embarked upon in this volume, following Doreen Massey's work on some philosophical assumptions that have stultified our understanding of the world (2004, p. 9), we approach the notion of children and globalization in three ways: (i) that we understand children as the product of interrelations constituted through spatial interactions, from the immensity of the global to the intimacy of the embodied, and that relations are necessarily embedded material practices, (ii) that we understand space as the sphere of possible multiple stories of becoming-other. That a post-developmental future is a multiplicity of children's stories-so-far in a plurality of spaces and across all scales, and (iii) that we understand that space and children are always already under co-construction, and that there is never closure. There are multiple possible futures that are not presaged by current neoliberal, academic or policy projects.

Of course, the various contributions of the book look upon globalization and childhood differently. Thus, the significance of globalization and how we understand it may vary according to what aspects of children's lives and situations are focused upon. Children may be directly involved in the neo-liberal (economic) project, with positive or negative outcomes. They may also be marginalized and exploited (socially, culturally and economically). Hence, what we understand as globalization (as with children and childhood) takes into account its multiple meanings and dimensions.

Geography and history, and how they come together in the notion of development, must be open in order to create a radical politics through which young people can become other than what geographic and historical frames suggest. What if imaginations are opened up to the possibility of multiple developmental trajectories for multiple storied young people? This question is the crux of *why children? why now?*

## Note

1. For example, a special issue of the *Transaction of the Institute of British Geographers* (2004) on globalization focuses on economics, cosmopolitanism, commodification and so forth, omitting any mention of children or young people.

## References

Aitken, S.C. (2001) *Geographies of Young People: The Morally Contested Spaces of Identity*, New York and London: Routledge.

Ansell, Nicola (2005) *Children, Youth and Development*, New York and London: Routledge.

Boundas, C.V. (1996) Deleuze-Bergson: an ontology of the virtual, in P. Patton (ed.) *Deleuze: A Critical Reader*, Oxford: Blackwell, 80–106.

Burman, Erica (1994) *Deconstructing Developmental Psychology*, London and New York: Routledge.

Deleuze Gilles and Felix Guattari (1988) *A Thousand Plateaus*, London: Althone Press.

Dyson, Tim (2001) A partial theory of world development: the neglected role of the demographic transition model in the shaping of modern society, *International Journal of Population Geography*, 7(2), 67–90.

Escobar, Arturo (1994) *Encountering Development: The Making and the Unmaking of the Third World*, Princeton: Princeton University Press.

Escobar, Arturo (2001) Culture sits in places. Reflections on globalism and subaltern strategies of globalization, *Political Geography*, 20, 139–74.

*Environment and Planning A* (2001) Theme issue entitled 'From Crib to Campus', Vol. 32(4).

Goddard, Jim, McNamee, Sally, James, Adrian and James, Allison (eds) (2005) *The Politics of Childhood: International Perspectives, Contemporary Developments*, London: Palgrave Macmillan.

Harvey, David (2005) *A Brief History of Neoliberalism*, Oxford: Oxford University Press.

Holloway, Sarah L. and Valentine, Gill (2000) Spatiality and the new social studies of childhood, *Sociology*, 34, 763–83.

Katz, Cindi (2004) *Growing Up Global: Economic Restructuring and Children's Everyday Lives*, University of Minnesota Press.

Laclau, Ernesto and Mouffe, Chantal (2001) *Hegemony and Socialist Strategy*, 2nd edition, London: Verso.

Massey, Doreen (2004) *For Space*, London: Sage.

Piaget, J. and Inhelder, B. (1956) *The Child's Conception of Space*. Translated by F.J. Langdon and J.L. Lunzer, London: Routledge & Kegan Paul Ltd (Original French Edition, 1948).

Pieterse, Jan N. (2001) *Development Theory—Deconstructions/Reconstructions*, London: Sage.

Ruddick, Sue (2003) The politics of aging: globalization and the restructuring of youth and childhood, *Antipode* 35(2), 334–62.

Teitelbaum, Michael, S. (1975) Relevance of the demographic transition theory for developing countries, *Science*, New Series 188, 420–25.

*Transactions of the Institute of British Geographers* (2004) Theme issue entitled 'Geography: Making a Difference in a Globalized World', 29(2).

Walkerdine, Valerie (1984) Developmental psychology and child-centered pedagogy: the insertion of Piaget into early education, in Julian Enriques, Wendy Hollway, Cathy Urwin, Couze Venn and Valerie Walkerdine (eds) *Changing the Subject: Psychology, Social Regulation and Subjectivity*, London and New York: Methuen, 153–202.

Walkerdine, Valerie (1988) *The Mastery of Reason: Cognitive Development and the Production of Rationality*, London and New York: Routledge.

Wallerstein, Immanuel (1980) *The Modern World System I: Capitalist Agriculture and the Origins of the European World-Economy in the Sixteenth Century (Studies in Social Discontinuity)*, London and New York: Academic Press.

Willis, Kate (2005) *Theories and Practices of Development*, London: Routledge.

# Reflections of Primitivism: Development, Progress and Civilization in Imperial America, 1898–1914

ELIZABETH A. GAGEN
*Department of Geography, University of Hull, Cottingham Road, Hull HU6 7RX, UK. E-mail: e.gagen@hull.ac.uk*

## Introduction

On February 15, 1898, an explosion sunk the *U.S.S. Maine* as it was harboured in Havana on a courtesy visit to Cuba. At that time, Spain still hung tenuously onto its last remaining Caribbean colony but with hostility growing against Spanish rule from within and increasing interest from the proximate US coastline, Spain's grasp was weakening. Most of the crew aboard the Maine were killed, but despite the fact that it was commonly believed to be an accident, the war hungry media on the US mainland whipped up a frenzy of speculation about Spanish involvement. In the following months an anaemic diplomatic effort failed and President McKinley declared war on Spain. It was a short and successful war for the Americans. In a matter of months, America secured a qualified independence for Cuba and in the process claimed Puerto Rico, Guam, Hawaii and the Philippines as its protectorates, launching itself, somewhat belatedly, as a New World imperial power (Brogan, 1999).

For many historians, the Spanish–American War marks the beginning of the Progressive Era, a period in US history marked by supreme faith in progress, self-confident, pious, and optimistic domestic reform and a newly invigorated foreign policy. In this paper I examine two sets of ideas and practices which emerge from this context—one seemingly domestic and other resolutely foreign. On the domestic front, America was beset at this time with anxieties about the progress of civilization. The nation seemed under threat from within (by declining birth rate in its women and waning virility among its men), and without (from the dilution of American stock by soaring immigrant numbers). Into this equation, the new discipline of psychology projected a profile of human development which offered to solve the crisis of civil progress by suggesting a means of securing 'normal' development. If the fledgling population could be instructed in normal development, and indeed improved upon, then the future of the nation was protected. Simultaneously, on the back of its newly acquired imperial status America began a programme of foreign development, one aspect of which was the establishment of schools, physical education programmes and playgrounds abroad. While these at first appear to be unrelated, or even in conflict—since one was pursued in the interest of domestic improvement and the other in the interest of improving foreign nationals—I argue that they fundamentally interpenetrate.

My aim is to triangulate these episodes, building on Massey's claim that it is essential to acknowledge the 'simultaneity of stories' if we are to genuinely argue that space is a product of interrelations (Massey, 2005, p. 9). In doing so, it is important to stress that plotting interrelations necessarily implies that connections, effects and influences rarely operate in one direction; rather, there are overlapping spheres of influence which operate in multiple directions simultaneously. In the context of late nineteenth-century ideas about normal development, American civilization and overseas colonies, it is commonly understood that foreign policy is ultimately driven by domestic concerns. While this was no doubt the case, I argue that conceptions of foreignness also fundamentally impinged upon domestic ideas and practices, creating a set of interpenetrating stories which deserve geographical attention.

This can be seen most clearly in the founding statements of psychology which established for the first time a normal curve of human development (Rose, 1998). In doing so, psychologists drew from a cocktail of evolutionary biology, physiology and, arguably, the emerging field of international relations to theorise a new science of child development. In essence, psychologists proposed a theory of individual development which was linked to the long term evolution of human 'races'. In the necessary hierarchicization of 'racial' development, childhood emerged as universally inferior, regardless of race.

Children were placed in a remote evolutionary phase of development, somewhere prehistorical and archaic, from where they would 'evolve' into fully fledged adults. Like America's 'underdeveloped' colonies, children were theorised as having to traverse evolution from prehistory to modernity. We are used to thinking through the colonial project from the perspective of the imperial centre, where colonial territories are placed on an evolutionary hierarchy. We are less familiar, I suspect, with reflecting on the degree to which this evolutionary hierarchy inflects the West's own thinking about itself. With regard to early child development theories, notions of foreign underdevelopment which surfaced in America's entrance into imperialism informed its own thinking about human development. The notion of 'underdevelopment', then, existed both on a geographical and temporal plane, relegating not simply foreign places and peoples to a premodern state, but simultaneously placing all children—American, European and Non-Western—in a universal state of primitiveness.

This intervention builds on existing work in children's geographies and new social studies of childhood which seeks to overturn the powerful tendency to view children through the binary construct of civilized adult versus natural child (James *et al.*, 1998). Whether this means children are seen as inherently wild, devilishly evil, vulnerably innocent, or incapably noble, the construct does violence to children's competence and their right to be valued on their own terms (Aitken, 2001). My goal here is to bring some historical clarity to the process by which such constructs get mobilised and how they draw in and rely upon a variety of other narratives. Children's geographies have chiefly focussed on current settings for their empirical and theoretical interventions. That work which has engaged with children in an historical context tends to be catalogued first and foremost as cultural and historical geography rather than identifying directly with children's geographies (see for example, Ploszajska, 1994, 1998; Lorimer, 2003). What I hope to demonstrate with this paper is that an historical perspective can broaden our preoccupation with the present and indeed add new insights. As Ruddick (2003) has already shown, this can be particularly fruitful when it comes to tracing the embedded discourses of globalisation and theories of childhood and youth. Crucially, she argues 'we will be better informed to critique and challenge particular aspects of globalization if we understand how some of its constitutive elements are being "smuggled" in . . . in the guise of new discourses around youth and childhood' (Ruddick, 2003, p. 335). And while her concern is current global restructuring, she is able to explain its menacing operations, at least in part, by recourse to historical narratives of childhood and youth. Equally, two papers in this volume discuss how representations of children in discourses around poverty in Non-Western contexts persist in representing children as vulnerable and needy (Bosco, this volume; Skelton, this volume). By offering a long view on the relationship between Western childhood and colonial traditions I suggest that these manoeuvres implicate not only children in non-Western contexts, but are embedded too in visions of Western childhood. My aim is to provide a critical history of the present in which current concerns about the global reach of liberal developmentalism are brought into historical perspective.

## Infantilizing Colonies/Colonizing Infants?

Every vigorous race, however rude and undeveloped, is, like childhood, worthy of the maximum of reverence and care and study, and may become the chosen organ of a new dispensation of culture and civilization. Some of them now obscure may be the heirs of all we possess, and wield the ever-increasing resources of the world for good or evil somewhat perhaps according as we now influence their early plastic stages, for they are the world's children and adolescents. (Hall, 1904, p. 748)

One of the most common observations in literature about colonial relations has been that indigenous populations are invariably cast as childlike in their nature. This has been connected to an imperial manoeuvre in which colonised cultures are differentiated from the imperial centre according to time scale (Said, 1978; Fabian, 1983; McClintock, 1995). Here, colonised places are located in the past—in a 'backward', 'primitive' or 'ancient' place—where they remain, in a state of arrested development, untouched by western progress. In evolutionary terms, the inhabitants of such places are similarly relegated to the past, frozen in some remote primitive phase of evolution. Historically, such policies allowed colonisers to infantilize native populations, justified a range of paternalistic policies and surveillance strategies and legitimated the exploitation of native economies. Likewise, such thinking has been prevalent in more recent developments of a new global order in the form of Rostowian modernization theory. Here, practices like child labour have been normalised as part of an inevitable teleology in which child labour is a necessary phase of normal economic development (Roberts, 1998). In this construction, the developing world is lagging behind the more 'advanced' countries of the west, but it is assumed that it will eventually travel the same route.

While this discursive manoeuvre is undeniable both in its practice and in the pernicious effects it continues to have, I suggest an alternative approach for thinking about the relationship between the colonised world and constructions of childhood. Rather than dwelling on the construction of indigenous populations as child-like, I argue here that the invention of normal development in the late nineteenth century rested on an understanding of western bourgeois children as inherently primitive. This builds on Stoler's (1995) observation that the 'regulatory mechanisms' of the colonial state—orders, strategies, surveillances, discourses—were as, if not more, interested in disciplining and constructing Europeans as they were colonised peoples. If this is the case, it is incumbent upon us to turn the mirror back on ourselves, to reorient our perspective temporarily away from constructions of the colonised in order to examine the impact of colonial thinking on the domestic scene. As Stoler (1995, p. 97) states, this 'entails tracing discourses on morality and sexuality through empire and back to the making of the interior frontiers of European nation-states'. It is precisely this manoeuvre I wish to make here, but in the context of tracing the discourse on child science through empire building and back to the domestic frontier of US nation making.

At the turn of the twentieth century, the United States was embarking on its own imperial project, ranging from military occupation and formal pacification to the extensive networks of indirect rule via education, economic dependency, and social policy. Central to these policies was an implicit understanding of indigenous populations as variously uncivilised, savage and childlike. Imperial programmes embarked on a civilizing mission to improve the condition of native populations in ways which were not unlike the tactics of European colonialism. However, as Stoler (1995, p. 150) notes, 'this equation of children and primitive, of children and colonised savage was not operative in overtly racist, colonial discourse alone'. On the contrary, the construction of children within the imagined and territorial borders the United States was also characterised by an equation with lower order beings and simian stock. As theories of evolution began merging within the modern discipline of child psychology, previously figurative associations of the child with the primitive gained new ground. In other words, US colonial policy did not simply construct the natives of the Philippines, Puerto Rico or Guam as primitive, but western child development theories themselves constructed the American child as inherently primitive. In this sense, the cultivation of civilization was as much about the production of the domestic bourgeois self as it was about the civilization of colonial subjects.

In the remainder of the paper, I focus on a series of endeavours carried out as part of an expansionist foreign policy: the extension of sport and recreation programmes to foreign territories.[1] These programmes were implemented by philanthropic recreation agencies and affiliated individuals whose explicit mission was to civilise native populations and encourage self-government. Despite the temptation to interpret this gesture within a colonial hierarchy which places indigenous communities at the bottom of the evolutionary ladder, I argue that these imperial missions were as much about the progressive civilization of the US nation as they were about the development of foreign territories. Before examining the implementation of recreation programmes abroad, let me begin by presenting a discussion of the anxieties that preoccupied America with regard to its progressive improvement and the solutions provided by the science of human development.

## Civilization, Nervousness and the Development of the Normal Curve

In 1893, Chicago held the World's Columbian Exposition showcasing America's power and superiority to the world (Rydell, 1984). Five year's prior to its first formal overseas imperial conquest, it was evident that, in spirit, America's outlook had reoriented from a determinedly westward, nation-building perspective to an equally dogged imperial gaze (Domosh, 2002). The frontier had formally closed, the west was colonised, the Native Americans subdued, and the nation was confident of its position in the hierarchy of world civilizations. In effect, America had come of age. Throughout the exhibition halls and displays, America displayed its success culturally, economically and politically. In spite of its pre-imperial status, the exposition relied heavily on displaying the conquest of Native American populations, thereby underscoring white America's cultural superiority (*Ibid.*). Cabinets and tableaux presented native populations in a state of under-development, located in some distant past where civilization was a future fiction rather than a present reality. In their 'pre-civilized' state, gender relations were yet to be differentiated, technologies yet to develop, evolution yet to proceed.

Alongside this prehistory, America was presented as a glorious victor, triumphant in its position at the front of the race towards civilization. In these show cases of primitivism— what Barkan and Bush (1995) call 'imperialist circuses'—America basked in its successful progress from a remote past. Native populations may share the same geographical territory, but temporally, they belonged in a different place—other worldly and anachronistic. Sitting rather uncomfortably alongside this exuberant and self-satisfied imaginary, however, existed a parallel debate which was significantly less definitive about civilization.

Central to the discourse on civilization that infused the 1893 Chicago Exposition was a belief that in its most advanced state, gender relations were marked by an absolute differentiation (Bederman, 1995; Domosh, 2002). What distinguished savagery from civilization was adherence to absolute forms of masculinity and femininity. The labour of indigenous women, for example, signalled their primitivism, while men who failed to provide for their family were marked by a lack of masculine values which similarly signalled incomplete civilization. At the centre of successful civilization, then, was the successful realisation of gender norms. While public spectacles like the World's Columbian Exposition could confidently present a desirable rendition of such gender distinctions, elsewhere in the cultural milieu, progress was not so clear cut. Numerous commentators have remarked upon the so called crisis of masculinity looming over American society at the end of the nineteenth century, and it becomes clear that where gender failed, civilization was also seen as under threat (Rotundo; 1993; Bederman, 1995; Kimmell, 1996).

By the 1890s, America had developed from a primarily agrarian economy, into a successful industrialised nation. Modern commercial centres replaced the rural frontier as the crucible in which American manhood would be cultivated. While this offered countless opportunities to breed economically successful men, it also produced a more complex environment for the cultivation of clear cut norms. Earlier in the century, Victorian manliness had been characterised by self-restraint, honesty, self-reliance, duty and the protection of family (Mangan and Walvin, 1987). However, with the influx of working-class immigrants, a series of economic downturns which threatened the success of small businesses and the increasing colonisation of the workplace by women, manliness appeared to be under threat (Bederman, 1995). In addition, a new disease appeared to be striking at the heart of civilized manhood. As many commentators have noted, neurasthenia was named in the late 1800s as a disease which peculiarly attacked middle-class, urbane, otherwise successful, men (Rosenberg, 1962; Lutz, 1991). Burdened by modern living, in which mental fatigue far outstripped physical labour, men were suffering from nervous collapse, which, in the eyes of social reformers, medics and politicians, was placing manhood in very serious peril.

During the last decade of the nineteenth century, reformers became increasingly concerned that the economic and cultural benefits of industrial progress were being counteracted by the neurasthenic leaching of all the most vital aspects of masculine identity. The idea that this condition could ultimately threaten American civilization caught the attention of many, including the future president, Theodore Roosevelt. Roosevelt became one of many voices compelling the nation to counteract the effects of city living with more vigorous and robust forms of exercise. Not only were adult men losing vitality with their desk jobs and effete virtues, but women too were failing the nation by turning their backs on motherhood. Roosevelt, spurred on by the fear of 'race suicide' that would surely materialise if men failed to assert their vigour and women failed to reproduce healthy stock, encouraged a return to hunting and outdoor activities and led a wider revival of sporting culture. While aspects of this movement were aimed at adult men and women, the most popular elements were reserved for America's youth.[2] A variety of organizations emerged from this period: the YMCA, the Boy Scouts of America, the American Physical Education Association and the Playground Association of America. All shared a common goal: securing a robust future nation by reinvigorating America's youth and transforming juvenile immigrants into civilized, robust, Americans. Exercise, sport and athletics would ensure that America's youth would be inoculated against future weakness and protected against neurasthenia.

Civilizing America's children while simultaneously protecting against the debilitating aspects of urban civilization became the primary task of America's youth organisations at the turn of the twentieth century. Like the tableaux of the World's Exposition, this relied on a teleological logic: that uncivilized young children would develop into civilized adults. Rather than operating along an evolutionary time scale of centuries and millennia, however, this new project relied on claims made by emerging child development psychology that stated that the improvement of the 'race' was achievable over the course of an individual's life time.

The principal proponent of these theories was G. Stanley Hall, a pioneering psychologist and later president of Clark University. In 1904, he published *Adolescence: its psychology, and its relation to physiology, sociology, sex, crime, religion, and education* which was widely consumed in the US and Britain. As a proponent of genetic psychology that was firmly neo-Lamarckian, his theories were quickly dislodged when Darwin's, and more importantly, Mendel's, findings were popularised in the early twentieth century (see Grinder, 1967). At the end of the nineteenth century, however, as America launched itself on the world stage, Hall's ideas about childhood development were widely absorbed. His influence can be traced through countless individuals and organisations, including the Progressive education movement, Boys' Clubs, the Camp Fire Girls, the child study movement, the playground movement and campaign for physical education both at home and abroad (Schlossman, 1973; Cavallo, 1981; Bissell Brown, 1990; Gagen, 2000a,b, 2004).

What I argue here, however, is that rather than see Hall's theories refracted through a variety of domestic settings, it is also useful to examine how America's newly imagined racial/imperial hierarchy was equally present in his theories of child development. A brief examination of his main theories brings this clearly into focus.

In *Adolescence*, Hall outlined a theory of child development and genetic psychology that attempted to describe not simply what happens during childhood and adolescence, but to identify a normal curve of development. Other fields were interested in 'morbid psychology' (what would now be called abnormal psychology), including criminal anthropology, sexology, degeneracy theory, social reform and medical psychiatry (Lutz, 1991). These disciplines busied themselves with the task of investigating errors in human development which caused deviance.[3] Academic psychology, however, was a new discipline keen to establish itself as a field in search of general laws of mental functioning and human development. Rather than searching for deviance, it focused instead on a systematic description of normal development. Only by doing so, could errors in normal development be identified and interventions considered which might secure the most desirable and robust adult.

Like many around him, including George M. Beard and the founders of the play and physical education movement, Hall was concerned with the pernicious qualities of urban life. While he concurred with Beard that neurasthenia was one such effect, he submitted a more complex and far reaching explanation for defective development. Hall's theory drew from the principles of genetic psychology which proposed that the nervous system archived the evolutionary history of the human race. The mechanics of development were explained by Hall through the theory of recapitulation (see also Ross, 1972; Gould, 1996; Noon, 2005). Human development, from embryo to adult, proceeded through each stage of evolution in collapsed time. The foetus appeared in a pre-human amphibian state, and from birth, infant, child and youth repeated each ancestral phase of evolution in collapsed time. Until the age of eight, children were thought to be in their most archaic, remote phase as they were temporarily controlled by the genetic memory of earliest mammalian development. 'To the young child', Hall states (1904, vol I, p. 220), 'there is no gap between his soul and that of animals'. Because this is the most uncivilized phase, children ought not to enter formal schooling until it ends at eight. Instead, they should be indulged in free play and allowed to explore their environment. From eight to twelve, Hall classified as the juvenile stage, represented by a period of slower physical growth and relative stability, suggesting 'some remote, perhaps pigmoid, stage of human evolution' (Hall, 1904, vol. I, pp. ix–x). During this period, children exhibited their most savage tendencies, defined as tribal, predatory, fighting, and roving. There might still be echoes of primate memories in behaviour, for instance, children of this age 'go on all-fours, are expert climbers, and assume, as savages often do in their dances, many strikingly ape-like attitude and contortions, while often their physical features—jaw, teeth, length of arm, etc.—point in the same direction' (Hall, 1904, vol. I, p. 217).

It was only at adolescence, beginning somewhere between 12 and 14 and lasting well into the twenties, that children entered the most recent phase of human development. This, for Hall, was the most crucial moment in development because it was during this period that additions could be made to development. Up to this point, development was merely recapitulatory, but during adolescence, the opportunity to advance the status of the race was possible. It was up to educators and parents, then, to ensure that at adolescence, children were allowed the best and most effective environment to develop into advanced and civilized adults.

Hall's intervention in the debate about civilization and the American nation offered some hope to those who feared the feminising effects of modern civilization. He offered assurances that given the correct environment, children could be coaxed

through the repetitive phases of early development and progress to new levels of achieve-
ment during adolescent years. Crucially, however, this pathway was used to explain why
Anglo-Saxon races had progressed to higher forms of civilization than other races
(Bederman, 1995). For those 'savage races'—including both African Americans and
Non-Western peoples—the full extent of this opportunity was yet to be realised.
Instead, individuals belonging to these races repeated the early phases of development,
but since they had not yet proceeded through civilization, development was essentially ter-
minated at the equivalent of white adolescence. This allowed Hall to argue that 'most
savages in most respects are children, or, because of sexual maturity, more properly, adolescents of adult size' (Hall, 1904, vol. II, p. 649).

In this respect, Hall's work duplicates a long colonial tradition which positions indigenous development at a juvenile stage. In addition, however, it also implicates child psychology in contemporary notions of racial development. By locating childhood as the bridge
between an ancient animal world and an advanced human one,[4] American notions of child
development had direct implications both for colonial communities *and* for Western childhood. Anxieties about civilization, when allied with a New World foreign policy, produced a universalising perspective that locked indigenous communities and domestic
children into the same constellation of ideas.

## Race Development and Civilizing Foreign Soils

Recapitulation theory merged the development of the child and the race. It was far reaching in its implications, accounting for the supposed supremacy of white America while
simultaneously offering a method for its betterment. Equally, in the context of late
nineteenth-century imperial ventures and a rapidly expanding immigrant population,
Hall submitted a theory that distinguished Anglo-Saxon from non-Anglo-Saxon children.
However, unlike European colonial discourses which refused the possibility of civilizing
indigenous populations, Hall stated that with the correct training and improvements,
'savage' races could grow into civilized races. This was a key twist in the logic of
improvement and rested on Lamarckian inheritance theory. Relying on the fact that individual improvements over the course of childhood could be transformed into inheritable
characteristics, Hall stated, 'a century with a race is no more than a year with an individual' (Hall, 1904, vol. II, p. 665). In other words, if individuals could be educated and
trained they could bypass centuries of racial stagnation. The possibilities of this belief
were endless, particularly in the context of America's newly acquired territories.

US foreign policy was infused with a New World authority which encouraged colonised
peoples towards self-government, as long as it conformed to an American norm.[5] Moreover, a theory of child development which connected savagery in western youth with
the evolutionary potential of world cultures provided a truly missionary goal for imperial
America. Speaking of the future of the colonies, Hall states, 'we must remember that their
aspirations are legitimate and that 'struggling nationalities are the jewels of history, and
the hope and promise of the world'' (Hall, 1904, vol. II, p. 665). Over the course of the
first decade of the twentieth century, Hall's thinking about child development moved
more firmly into the realm of racial improvement. In 1910, Hall and George Blakeslee,
a historian and pioneer of International Relations at Clark University, founded the
*Journal of Race Development* which later evolved into *The Journal of International
Relations* and is still published today as *Foreign Affairs* (Blatt, 2004). The *Journal of
Race Development*, extended what was already prominent in Hall's theories: the belief
that using scientific knowledge the 'inferior races' of the world could be improved in
service of a more efficient and economically liberal world. In effect, the project united

the disciplines of evolutionary biology, psychology and international relations in order to outline applied theories of racial improvement. What is often overlooked, however, are the latent implications of these ideas for western childhood.

In plotting the racial desires of contributors over the journal's early years, Blatt (2004) observes that Progressive Era reformers and civil servants were keen to differentiate American imperial policy from old world colonialism. Crucially, the difference lay in the potential improvement of native populations. According to the American writers of the *Journal of Race Development*, where European colonialism had seen enduring racial hierarchy, the US saw opportunities for progress and improvement. In the opening editorial, Blakeslee argues that US imperialism was 'not about how weaker races may best be exploited, but how they may best be helped to be stronger' (Blakeslee, quoted by Blatt, 2004, p. 694). America's new colonies were to provide experimental testing grounds for intervention and positive change, where colonies were to represent 'flagships in a developmental project' (Blatt, 2004, p. 697). This is most plainly stated in a speech reprinted in the Journal by the Commissioner of the US Civil Service in the Philippines, in which he writes: 'The reform movement is dominant now in America, insisting not only on clean governmental operations, but also the enactment of laws for the betterment of the people, for their moral, mental, and physical elevation. [It is] desirable and essential that these reform influences and the power of public opinion in the United States extend to the Philippines. [This will give] the Filipino people every opportunity for development' (*Ibid.*, p. 697).

The question was how? How would this improvement take place? Through what technologies of development? This is where Hall's theoretical work proved useful. Not only had he argued vociferously that an entire race could be improved over the course of one generation if children could be coaxed through normal development and adolescents could be suitably augmented, but his students and supporters had launched a widespread domestic campaign to implement these ideas in schools and playground across America (Gagen, 2000a, 2004, 2006). Child rearing manuals, pedagogical theories, recreation facilities, playground and physical education provisions were all influenced by Hall's developmental thinking. Physical development became allied with the successful realisation of normal development and consequent racial improvement. Indeed, the popularity of these ideas was also bolstered by the fact that anxieties about neurasthenia also prescribed physical exercise as a cure. Physical improvement thereby became a powerful mechanism for guaranteeing normal development and racial progress. Playgrounds and physical education curricula developed across the United States in an effort to improve immigrant bodies and ensure the strength of future citizens. Moreover, in the context of international development, this technology proved immeasurably popular and had the support of the muscular President Roosevelt.

For many commentators, Roosevelt's approach to foreign policy was indicative of his attitudes toward civilization, race and masculinity. In particular, Thomas Lutz (1991) connects Roosevelt's fears about his own neurasthenic condition with his political views on foreign policy and his vigorous support for sport and leisure among the US male population. Like the writers in the *Journal of Race Development*, Roosevelt believed both in the legitimacy of US imperialism and the implied liberal developmentalism. Specifically, he was a strong supporter of physical improvement as a route to self-government—the ultimate goal of foreign development. For Lutz (1991, p. 80), 'Roosevelt explicitly related the fitness that results from the strenuous life to people's fitness to rule themselves'. Therefore, when volunteers from the Playground and Recreation Association of America (PRAA) and the YMCA offered to launch programmes overseas, both in US colonies and in non-US territories, Roosevelt was enthusiastic.

During the first two decades of the twentieth century, American agencies dispatched volunteers and workers across the colonised world to establish schools, playgrounds and recreation programmes. The work was seen as more of an extension of domestic projects rather than a comprehensive and independent programme. The journal of the PRAA, entitled *The Playground*, included regular features on its foreign initiatives, entitled 'Notes from Abroad'. On these pages, overseas workers reported on the education and recreation works being carried out in places like Tahiti, Uruguay, Puerto Rico, the Philippines, India, Hawaii, China, and Panama. The reports speak optimistically of the improvement of indigenous children and the transformative possibilities of physical culture and play. In Puerto Rico, for example, the Supervisor of Athletics and Playgrounds in San Juan writes, 'Although climate, tradition and temperament have combined to retard physical training and playground activity in Porto Rico in the past, the systematic efforts inaugurated in 1913 have produced surprisingly rich results thus far. In fact, in many instances, the interest and ability exhibited may bear favorable comparison with physical education in other lands where material conditions are far better' (Wiggins, 1915, p. 319). What becomes clear in these discussions is that while local populations were considered to have been 'underdeveloped' in the past due to inherent environmental and evolutionary obstacles, these were believed to be surmountable problems. Unlike European colonials who saw their subjects as locked in an antique state, American colonisers considered uplift and improvement to be an entirely feasible goal.

Such improvements would be achieved through the encouragement of proper developmental phases and correct physical training. Advising trainee supervisors on the pages of *The Playground* journal, Henry Curtis writes, 'Play is the motor habit and spirit of the past race persisting in the present' (Curtis quoted by Hermann, Richards, Chase and Rowley, 1914, p. 65). Adhering closely to Hall's theories of child development he writes that if children, both at home and overseas, are encouraged to repeat the correct phases, they will surely develop into self-governing adults. Demonstrating this teleology, he explains how children could 'evolve' from prehistory to modernity: from primate to human adult. Through a series of rhetorical questions, he asks: 'What made the monkey's chest so deep?—his climbing. What makes his waist so small?—his vaulting. What makes his loins so strong?—his balancing on shaking branches. What has developed man's courage, judgement and vision, self-reliance and confidence?' (*Ibid.*) His answer, of course, was team sports and play. If the populations of developing countries and colonies could be instructed in the correct developmental sequence of play and exercise, they too would develop into modern self-governing citizens.

An example from India is particularly instructive. After founding California's first public playground, the philanthropist Charles M. Goethe visited India and was struck by the possibilities for improving its native children. In an article for *The Survey* in 1916, he uses a story about his encounter with a young, lower-caste Indian boy to illustrate how playgrounds could provide a catalyst towards civil advancement. He begins by describing his first encounter with the young child. At this point, the child was hungry, emaciated, unclean and disempowered. Faint with hunger he passes out and is taken in by two Americans teaching in an American school in Calcutta. Here the child is exposed to American play and education, and from that point on 'he grew much like the boys on an American playground' (Goethe, 1916, p. 113). The boy grows up to be a successful college athlete; he attends a European university and returns to India to teach in the same school that improved him. Alongside the other Americans who continue to run the School, he helps transform malleable young Indian boys 'who are making a New India' (*Ibid.*). It is clear from this narrative, that the metamorphosis is possible because of the interventions of American supervision. As Goethe (1916, p. 113) states, 'To this great future, the playground movement is contributing its share'.

The success of the young boy's transformation persuaded Goethe to sponsor an additional three playgrounds in Calcutta. The following year, he arranged for two recreation workers to be transferred from Massachusetts and the playgrounds, like the one pictured below, began their work on young Indian children (Figure 1).

Some of the reports featured in *The Playground* journal repeat familiar colonial constructions. In the report from Tahiti for example, the Supervisor of Public School Athletics in San Francisco describes the island as a 'land where nobody works' (Peixotto, 1914, p. 50). Despite this, however, she is optimistic that infinitesimal markers of evolutionary potential offer America hopeful avenues for intervention.

Speaking of the Tahitians, she writes, 'Even though they reduce their long leisure hours by copious sleep, still, like other "uman animals" Tahitians have their "lay instinct"' (Peixotto, 1914, p. 50). As long as these are present, recreation workers believed they could be shaped into citizenly qualities.

In all these 'Notes from abroad', native children of Tahiti, India and elsewhere were exoticized and primitivised, constructed as pre-human, pre-civilised, but spilling over with the potential to become modern democratic subjects. In this sense, colonial children were constructed in ways which were compatible with the American children in Hall's theories. In Hall's recapitulation theory, every child—American, Filipino, or Indian—was in a state of archaic underdevelopment, and all had the potential to become modern. Alongside the articles on play in foreign lands, Hall's wife, Katherine Stanley Hall, wrote a piece entitled 'Children at play in many lands' in which she argued that 'the instinct of play seems to be about the same in all lands' (Hall, 1915, p. 323). This was the cornerstone of American child development. It justified the belief that American children were miniature savages requiring careful tutelage for proper development; and it justified colonial education programmes where future self-government, as America conceived it, was at stake. This produced an uncanny parallel between American childhood in which the young were cast as prehistoric time travellers on their way towards modernity (see Figure 2) and the children of America's

**Figure 1.** Goethe's playground, Calcutta, *The Survey*, November 1916.

The Cave Men

**Figure 2.** American children simulating 'The Cave Men', *The Playground*, 1911, volume V, number 1.

colonies whose 'savage' state did not need to be re-enacted by dramatic play, but rather, passed through and augmented in a more urgent passage to a similarly modern place.

## Conclusion

My aim in this paper has been to demonstrate that the significance of constructing native populations as 'primitive' sometimes masks an equally significant manoeuvre to construct western childhood as equally prehistoric, and indeed, barely human. In the context of imperial America, the scientific technologies circulating in early academic psychology were deployed in the service of colonial ventures, and equally, colonial constructions of racial difference underpinned domestic theories of child development. There were strategic reasons why this slippage was politically useful. Primarily, the arrival of millions of immigrants caused deep anxiety among America's Anglo-Protestant elite. Fears about 'race suicide' and the degeneration of America from within provided a context in which a more optimistic theory of evolution was welcome. Old World notions of racial hierarchy did not offer much scope for transforming 'inferior' stock into successful citizens. If America was to succeed in its millennial goal and spread that supremacy around the world, it needed a more optimistic theory of racial improvement. Hall's theory of recapitulation offered just that. In opposition to European colonial discourse which froze colonised peoples in an ancient moment, Hall collapsed time and space into childhood. Instead of locking indigenous populations into prehistory, he relegated Western children to the same archaic place. The difference being that development was no longer immobilised but existed, both for America's colonised and its domestic youth, as thoroughly plastic. Among other things, this has produced a tenacious condition of perpetual becoming for Western childhood, while maintaining America's commitment to liberal developmenalism in an intensifying global order.

## Acknowledgements

Thanks to Stuart Aitken, Anne Trine Kjørholt and Ragnhild Lund for organising the workshop on *The Global Child* for which this paper was first written. Thanks also to the other participants whose ideas and contributions have helped me develop this paper immensely. Further thanks to Stuart for organising these theme issues and, along with anonymous referees, for providing helpful comments on an earlier draft of this paper.

## Notes

1.  Details of these programmes were recorded in the journals of a variety of social reform agencies. The chief organisation coordinating recreation both in the US and abroad was the Playground Association of America, later the Playground and Recreation Association of America. Their monthly journal—entitled The Playground—published regular updates on their foreign programmes. This paper draws on these records, along with other contemporary publications on child development psychology.
2.  Women were considered equally responsible for improving physical fitness as it was understood that unless they themselves were physically robust, they would not be able to breed healthy offspring. This relies on Lamarckian evolution which is explained later in the text.
3.  As Lutz also argues, it was not until Freud that we see the proposal that abnormal psychology is not a degenerate form of normal psychology, but an exaggeration of it.
4.  Noon (2005) takes this further, arguing that recapitulation theory did not simply place children at an earlier developmental stage, but theorised them as qualitatively different, non-human creatures. My argument here is that this not only legitimated racially informed colonial policies, but also animalised and primitivised Western childhood.
5.  Natually, this was not an altruistic venture. US colonies were cultivated as captive markets for American goods, just as European colonies had. Unlike European colonies, however, US protectorates were actively encouraged towards self-government in the service of 'liberal developmentalism'. Central to this agenda was the belief that foreign nations ought to duplicate America's own developmental trajectory with a commitment to free trade and market values. Indeed, Neil Smith argues that this period can be seen as a turning point in the history of imperial expansion, lying between a period when economic expansion required territorial gain, and an era in which economic expansion required, not territory, but a functionally related globalised world. This short-lived imperial moment was the last gasp of an expansionist philosophy whereby growth was achievable via land grabs. In this equation, America's appropriation of the 'geographical crumbs' of the Spanish Empire represents the beginning of a much more significant project: the century long (and indeed on-going) commitment to building an American Empire at the helm of a globalised world order (Rosenberg, 1982; Smith, 2003, p. 16).

## References

Aitken, S. C. (2001) *Geographies of Young People: The Morally Contested Spaces of Identity*, London and New York: Routledge.

Barkan, Elazar and Bush, Ronald (1995) *Prehistories of the Future: The Primitivist Project and the Culture of Modernism*, Stanford, CA: Stanford University Press.

Bederman, Gail (1995) *Manliness and Civilization: A Cultural History of Gender and Race in the United States, 1880–1917*, Chicago, IL: University of Chicago Press.

Blatt, Jessica (2004) To bring out the best that is in their blood: race, reform and civilization in the Journal of Race Development, 1910–1919, *Ethnic and Racial Studies*, 27(5), 691–709.

Brown, Victoria Bissell (1990) The fear of feminization: Los Angeles high schools in the Progressive Era, *Feminist Studies*, 16(3), 493–518.

Brogan, Hugh (1999) *The Penguin History of the USA*, London: Penguin.

Domosh, Mona (2002) A civilized commerce: gender, race and empire at the 1893 Chicago Exposition, *Cultural Geographies*, 9, 181–201.

Cavallo, D. (1981) *Muscles and Morals: Organized Playgrounds and Urban Reform, 1880–1920*, Philadelphia, PA: University of Pennsylvania Press.

Fabian, Johannes (1983) *Time and the Other: How Anthropology Makes its Object*, New York: Columbia University Press.

Gagen, Elizabeth A. (2000a) 'An example to us all': child development and identity construction in early twentieth century playgrounds, *Environment and Planning A*, 32(4), 599–616.

Gagen, Elizabeth A. (2000b), Playing the part: performing gender in America's playgrounds, in S.L. Holloway and G. Valentine (eds) *Children's Geographies: Playing, Living, Learning*, London: Routledge, 213–29.

Gagen, Elizabeth A. (2004) Making America flesh: physicality and nationhood in early twentieth-century physical education reform, *Cultural Geographies*, 11, 417–42.

Gagen, Elizabeth A. (2006) Measuring the soul: psychological technologies and the production of physical health in Progressive Era America, *Environment and Planning D: Society and Space*, 24(6), 827–49.

Goethe, C.M. (1916) Nirman Singh: The story of a little lizard eater, who stands for the childhood needs of the unnumbered children of British India', *The Survey*, November, 4, 111–118.

Gould, S.J. (1996) *The Mismeasure of Man*, New York: W.W. Norton and Company.

Grinder, R. (1967) *A History of Genetic Psychology: The First Science of Human Development*, New York: John Wiley.

Hall, Granville Stanley (1904) *Adolescence: It's Psychology and its Relations to Physiology, Anthropology, Sociology, Sex, Crime, Religion and Education*, New York: Appleton.

Hall, Katherine Stanley (1915) Children at play in many lands, *The Playground*, IX, 9, 323–7.

Hermann, E., Richards, J.R., Curtis, Chase, J.H. and Rowley, L.E. (1914) Discussion of Dr. Curtis' article, *The Playground*, VIII, 2, 65–75.

James, A., Jenks, C. and Prout, A. (1998) *Theorising Childhood*, New York: Teachers' College Press.

Kimmel, Michael (1996) *Manhood in America: A Cultural History*, New York: The Free Press.

Lorimer, H. (2003) Telling small stories: spaces of knowledge and the practice of geography, *Transactions of the Institute of British Geography*, 28, 197–217.

Lutz, Thomas (1991) *American Nervousness, 1903*, Ithaca, NY: Cornell University Press.

Mangan, J.A. and Walvin, James (eds) (1987) *Manliness and Morality: Middle-class Masculinity in Britain and America, 1800–1940*, Manchester: University of Manchester Press.

Massey, Doreen (2005) *For Space*, London: Sage.

McClintock, Anne (1995) *Imperial Leather: Race, Gender and Sexuality in the Colonial Conquest*, New York: Routledge.

Noon, D.N. (2005) The evolution of beasts and babies: recapitulation, instinct, and the early discourse of child development, *Journal of the History of Behavioural Sciences*, 41(4), 367–86.

Peixotto, E. (1914) Recreation sketches in the South Seas, *The Playground*, VIII, 2, 50–5.

Ploszajska, T. (1994) Moral landscapes and manipulated spaces: gender, class, and space in Victorian reformatory schools, *Journal of Historical Geography*, 20(4), 413–29.

Ploszajska, T. (1998) Down to earth? Geography fieldwork in English schools, 1870–1944, *Environment and Planning D: Society and Space*, 16, 757–74.

Roberts, Sue (1998) Commentary: what about the children?, *Environment and Planning A*, 30, 3–11.

Rose, Nikolas (1998) *Inventing Ourselves: Psychology, Power and Personhood*, Cambridge: Cambridge University Press.

Rosenberg, C. (1962) The place of George M. Beard in nineteenth-century psychiatry, *Bulletin of the History of Medicine*, 36, 245–59.

Rosenberg, R. (1982) *Spreading the American Dream: American Economic and Cultural Expansion, 1890–1945*, New York: Hill and Wang.

Ross, D. (1972) *Stanley Hall: The Psychologist as Prophet*, Chicago, IL: University of Chicago Press.

Rotundo, Anthony E. (1993) *American Manhood: Transformations from the Revolution to the Present Era*, New York: Basic Books.

Ruddick, Sue (2003) The politics of aging: globalization and the restructuring of youth and childhood, *Antipode*, 35(2), 334–62.

Rydell, Robert W. (1984) *All the World's a Fair: Visions of Empire at American International Expositions, 1876–1916*, Chicago, IL: University of Chicago Press.

Said, Edward (1978) *Orientalism*, London: Routledge.

Schlossman S. (1973) G. Stanley Hall and the Boys' Club: conservative applications of recapitulation theory, *Journal of the History of Behavioral Sciences*, 9, 140–47.

Smith, Neil (2003) *American Empire: Roosevelt's Geographer and the Prelude to Globalization*, Berkeley and Los Angeles, CA: University of California Press.

Stoler, Ann Laura (1995) *Race and the Education of Desire: Foucault's History of Sexuality and the Colonial Order of Things*, Durham, NC: Duke University Press.

Wiggins, B.E. (1915) Playgrounds and physical training in Porto Rico, *The Playground*, IX, 9, 319–23.

# Childhood as a Symbolic Space: Searching for Authentic Voices in the Era of Globalisation

ANNE TRINE KJØRHOLT
*Norwegian Centre for Child Research, Norwegian University of Science and Technology, 7491 Trondheim, Norway. E-mail: anne.trine@svt.ntnu.no*

## Introduction

The UN Convention on the Rights of the Child can be seen as an aspect of globalisation processes, since it presents particular images of children as subjects that are claimed to be universal. During the last 15 years, in addition, we have witnessed an increasing emphasis on 'children as participants' in society, including an emphasis on children as political activists in different parts of the world. Rights to active social participation, which are often interpreted as rights to citizenship, have been described as revolutionary compared to earlier international declarations of children's rights and as a step towards increased recognition and justice for a group of human beings who are typically marginalized in society. However, the emphasis on children as claimers to rights to participation, as set out in the UN Convention, is different from the idea of and forms of participation relating to children as social actors in everyday life, as unfolded in families, kindergartens, schools, local communities and society per se.

Children are social participants in societies and cultural life in many respects, as, for example, labourers, soldiers and consumers, and they reproduce and produce culture in everyday lives in different localities on a par with adults. They are co-constructors of their childhoods and active agents in establishing relationships with adults, as well as with other children. They are caring subjects and embodied beings who contribute emotionally to their own and others' welfare and quality of life. Human life itself presupposes participation. Studies of infants show that children have an innate ability from birth actively to influence communication with other people in their surroundings (Threvarthen, 1973; Bråten, 1998). However, the shape and expression that such participation is given, as well as the domains in which children are granted an opportunity to participate, are part of an intergenerational social order and are deeply embedded in cultural notions of what it means to be a child. More than that, children's different forms of participation are an integral part of social life in particular places, covering a range of different social practices, relationships and meaning-making processes in different localities, and producing and reproducing economic structures and culture (Katz, 2004). As Tatek (this volume) describes, children in the global south contribute immensely to the survival strategies of the households they belong to by taking part in different income-generating activities, often without the involvement of adults.

However, these forms of children's active participation in the economic and social reproduction of society are often ignored, partly due to the dominance of developmental discourses, which have been highly criticised for producing images of children as vulnerable, immature and in need of education and socialisation if they are to develop into fully competent citizens (Woodhead, 1999). The complex ways in which children represent an integral part in processes of economic, political and cultural reproduction in a society are thus to a great degree overlooked. Paradoxically, contemporary global discourses on children's rights to social participation, which emphasise children's autonomy and competence, do not address these forms of participation. A pertinent question is whether rights discourses therefore have the unintended effect of contributing to and accelerating processes of invisibility, detecting children's traditional roles in social reproduction in different societies and localities. Furthermore, this fundamental question relates to a further enquiry as to whether or not there are potential conflicts and tensions between notions of participation as constituted within rights discourses that are claimed to be global, and children's 'traditional' and integrative forms of participation in social, political and economic life in different local and national contexts.

In her book *Growing Up Global*, based on extensive fieldwork with children in Sudan and New York, Cindi Katz reveals how processes of development and global change have had similarly disintegrative effects on communities in the global South and the global North (Katz, 2004). Empirical analysis is needed to reveal the complexities and dynamic interrelationships between global changes and local practices that are developed by children as active social participants (Hengst, 2006).

Nevertheless, contemporary discourses relating to the UN Convention on the Rights of the Child address political participation by and the citizenship of children in new ways. Not only do the participation rights listed in the CRC represent a tool for the potential of the individual child to have an increased voice in matters that affect her/his life; the CRC also opens up space for new and sophisticated forms of child participation in social and political reproduction, and on both the national and international levels. Rights discourses representing changes in the ways in which children are expected to take part in the social and political reproduction of contemporary societies highlight the need to investigate the changing discourses critically, to examine implementation of the UN Convention in different countries, and to consider how it is affecting children's livelihoods in different localities.

The aim of this paper is therefore to discuss the close interplay between global discourses on children as citizens, notions of (a good) childhood at the national and local levels, and discourses on nationality and democracy. This dynamic relationship highlights on the one hand the significance of children as central social actors, and on the other hand how social constructions of children and childhood are closely intertwined with economic, cultural and political transformations in society. The discussion is empirically grounded in the analysis of participatory projects in two Nordic countries, initiated in the early 1990s, 'Try Yourself' in Norway, and 'Children as Fellow Citizens' in Denmark, in both of which the constitution of children as autonomous, authentic social actors is central. The main argument I present is that, to an increasing degree, childhood is constructed as a symbolic space reproducing symbolic values related to democracy, national identity, autonomy and authenticity. In this way, children are constituted as social participants, actively contributing to reproducing culture and national identities in new ways.

## Mary: Labourer and Political Actor

There is a major focus on the constitution of children as authentic, autonomous participants in society in international society. As part of my discussions of 'childhood as a symbolic space', I shall present the voice of Mary Benjamin Olarita, a 14-year-old girl from the Philippines whom I met at the international conference on 'Children's Rights and Wrongs', arranged by the Centre for World Dialogue in Nicosia, Cyprus, on the tenth anniversary of the UN Convention on the Rights of the Child in 1999. Mary gave a moving testimony about her life. For many years she had been working as a labourer in the docks in the slums of Davao City from seven o'clock in the evening to seven o'clock in the morning. Her job was to load 50-kg sacks on and off barges. The wages were close to minimal, and if she ate anything during the 20-minute break, they would be reduced accordingly. Most of the money Mary earned was given to her family, which depended on her wages for their living. The rest was used to pay for Mary's schooling. Mary slept three hours a day, from eight to eleven in the morning, before attending school until about five in the evening. Often she fell asleep at school because she was so tired.

Besides being a child labourer, Mary was also a child activist. She participated in the Global March of the mid-1990s, and, together with the other children who were 'child-speakers' at the conference, sang one of the songs that were written for the March, entitled 'Stop child labour!' The message of the children's voices was sung in a manner that was both emotional and confrontational. Mary was quite professional in her role as a child activist, as she was used to testifying about her hard life and speaking out against child labour at large conferences, like that in Nicosia, which attracted 450 participants from all over the world. However, at the end of her prepared speech she became lost for words, and tears replaced her appeal to 'Stop child labour'! With tears streaming from her eyes, she cried out: 'But how can I speak up in support of stopping child labour? My family needs the money I earn! They cannot afford to pay for my school!'

Mary's testimony clearly illustrates certain dilemmas and ambiguities embedded in discourses on children's rights that are also of great relevance for forms of children's participation other than child labour. Her voice demonstrates the global character of discourses on children's rights in their having a political voice, as well as certain paradoxes associated with the construction of the autonomous, participating child subject that are embedded in rights discourses. Mary is given the right to participate and express her opinions about her living conditions, as stated in the UN Convention on the Rights of the Child. However, although her voice in support of 'stopping child labour' may be heard, this simply leads us to question the consequences for her own and her family's lives. Without fundamental changes in the world economy, as well as in working and living conditions in Davao City, Mary's life will probably change for the worse. Her participation in child labour represents a ticket for her entry into school, as well as to achieving a better standard of living for her family.

The voices of Mary and others at the conference vividly illustrate that children today are actively contributing in the reproduction of society in different ways, economically, politically and culturally. The message of the global march: 'Stop child labour', was extremely powerful because it was expressed in a moving way by the voices of the children themselves, constituted as political actors with equal rights as adults. On the other hand, the children were speaking against other forms of child participation, such as child labour. Mary's spontaneous emotional utterance at the end of the performance, contradicting the message in the prepared song, clearly reveals the paradoxes related to this political act, as well as tensions that arise between different forms of participation at the global and national levels. In international society, child activists like Mary can be seen as symbolic actors contributing to the production of images of a global society aiming at further development, progress and humanity. It has been argued that children are often used as symbolic participants more than empowered actors enacting real influence, and that participatory projects for children and young people might turn into 'prestige projects' serving as tokens for certain political decisions, rather than realising children's actual interests (Hart, 1992; Haugen, 1995).

### 'Children's own Culture': Searching for Authentic Voices

I now present two participatory projects from the Nordic countries to provide empirical evidence for my argument. To a great extent, political discourses in Norway and Denmark represent childhood as a separate age-segregated life-world in which children create 'their own culture' relatively independently of adults (Kjørholt, 2004). The participatory project, 'Try Yourself', initiated in the early 1990s by the Art Council in Norway, can be seen as a 'key project' that to a high degree represents core values related to notions of (a good) childhood in Norway. The aim of this project was to provide grants to children aged 7–14 to carry out their ideas and creative activities in their local communities. Eleven municipalities took part in the three-year project, in which children are constructed as competent actors belonging to a specific age group in need of freedom from adult control and power. Although they are seen as *different* from adults in respect of having their own peer culture, children were also recognized as being the *equals* of adults, as autonomous subjects with a right to influence society and to apply for funding for cultural activities without asking their parents or other adults for permission. More than 1700 children aged 6–15 took part in the project during the three-year project period. They received funding for a lot of different activities, such as running 'tinkering clubs' (boys of all ages—groups fiddling with cars), 'caring for cats' clubs, arranging photo exhibitions, theatre clubs, making newspapers and many other activities. They asked for funding to sell eggs, construct huts in tree tops, establish cafes, produce art for their animals and so on. The project opened up new positions for children as autonomous social participants in their local communities. The idea of the project can be seen as quite radical, freeing children in some respects from their positions as dependants in their families and on adults, and opening up new positions for children as social participants in public spaces (Kjørholt, 2001, 2004).

The 'Try Yourself' children in my case study placed themselves in different ways as autonomous competent subjects within discourses of children as social participants in society Their *autonomy*, however, was not constructed as a counterpart to *dependency*. Quite the contrary, the construction of identities as competent social participants derived from intertwined processes of autonomy and of belonging to various kinds of communities, intergenerational as well as age-related.

The notions of childhood as being threatened by adult control and intervention represents, I suggest, a powerful position in discourses concerning what is (a good) childhood in Norwegian society. The project 'Try Yourself' can be seen as aimed at revitalising a 'childhood of the past' or 'the authentic childhood' and realising 'children's need to be children', associated with play. To put it bluntly, children were paid to play. 'Try Yourself' can thus be interpreted as an exciting effort to construct new positions for children's social participation in their local communities by including them as citizens and recognising their particularity. On the other hand, in other respects children are not taken seriously as citizens on a par with adults, because they are placed in discourses that also refer to romantic notions of 'the playing, innocent child' living in their own separate lifeworlds within an age-segregated social order.

Related to this view is the conceptualisation of children as belonging to a group that practises its own authentic culture independently of adulthood, which is greatly emphasised within the Nordic countries, in both policy and research. In the Nordic countries, the concept of 'children's own culture' is used by ethnologists, together with researchers from other disciplines within human research (Enerstvedt, 1971; Mouritsen, 2002; Selmer-Olsen, 1990; Junker, 1998). This research tradition has to a great extent impacted on political decisions related to cultural and educational policy (Tingstad, 2003). From a discourse-theoretical perspective, the circulation of images of 'children as a group with their own authentic culture' in the fields of both child research and policy reveals the powerfulness of this particular construction. It has been argued that the use of the concept in policy can be traced back to the Ministry of Culture in Denmark in the latter part of the 1960s, a time when there was growing concern about the lack of interest in children as a target group in respect of mainstream cultural policy. The importance of rethinking cultural policy to include cultural events and activities made specifically for children, such as museums, literature, art, etc., was emphasised. This was seen as part of a movement addressing a need for the emancipation of children (Tingstad, 2006). As part of the discussion of the development of a particular cultural policy for children, it was argued that not only were they passive recipients, but also social actors contributing to producing their own culture (Tingstad, 2006). From the 1990s, notions of 'children's own culture', representing a particular understanding of childhood, have come to occupy an increasingly powerful position in the discursive field among both researchers and various professional groups (Kampmann, 2003). Actors in the field present this position as representing a shift from a focus on the *developing child* to the *competent child*, from *pedagogy* to *culture*, these being seen as opposites. The concept of 'children's geographies', emphasising the importance of recognising and studying children's places as particular sites for processes of cultural production and reproduction among children seen as competent social actors, can to some extent be seen as an analogue to the concept of children's own culture.

The notion of 'children's own culture', also called 'play culture' in Nordic policy and research, refers to children's own play and cultural activities, understood as an independent site with its own folklore and rituals (Opie and Opie, 1969; Junker, 1998; Mouritsen, 2002). Children are constructed as competent, creative actors, and childhood is thought of as a domain in which children's own authentic culture unfolds and can be understood relatively independently of adulthood. Researchers in this field have been concerned for the danger that traditional play culture made *by* children should disappear (Enerstvedt, 1971; referred in Tingstad, 2003). These researchers have been criticised for being tied to romantic notions of child cultural communities, for stressing the rupture between childhood and adulthood, and for overlooking continuity with and dependent relations on adulthood (Gullestad, 1994).

The particular construction of children as belonging to a group with 'their own authentic culture', and the construction of the 'competent, participating and autonomous child', which is prevalent in contemporary discourses in Norway and Denmark, represents a reconstruction of a 'childhood of the past'. 'Traditional' notions of a good childhood are closely related to play and the potential of children to move freely and practice cultural activities with their peers, in particular in outdoor spaces without adults controlling and organising their space. Nature is seen as a particularly preferred place for children's play and activities with their peers. Climbing trees and building huts are thus seen as core activities constituting 'the Norwegian' childhood. The increasing institutionalisation of children's lives, manifested in the expansion of day-care centres during the last two decades and an increased degree of adult control and organisation of children's time and space in their leisure time, is regarded as a threat to a 'natural and free childhood' in which children can develop their own activities together 'freely'. The project 'Try Yourself' thus represents an attempt to prevent an increasingly adult control of time and space in children's lives. Furthermore, it can be also seen as part of social reproduction processes reconstructing Norway as a modern participatory democracy embracing difference by being able to include another group of 'different citizens', that is, children. Before elaborating more on this point, I shall briefly present the Danish project.

## Autonomy and Self-determination as Ideology

The aim of the other project, initiated in Denmark and entitled 'Children as Fellow Citizens', was to empower children as citizens and increase their ability to influence their daily lives. As part of the project, various activities aimed at promoting children as social participants in different contexts were implemented in five local communities in Denmark. One of these activities dealt with children's rights to be active participants in day-care institutions. Children's ability to make decisions for themselves in an institutional context such as a day-care centre was a core issue.

The project, initiated in the 1990s by the Danish Ministry of Culture, reveals particular notions of toddlers and children below school age as 'fellow citizens' in day-care centres,[1] similar in some ways to constructions of children in Norway. These notions connote specific ideological and moral values. Based on an analysis of public texts related to this project, I have argued elsewhere that the construction of the child subject is related on the one hand to processes of individualisation,[2] and the construction of the autonomous, self-determining subject in late modern societies in the Western world in general, and on the other hand to particular cultural notions of 'the free child' that are current in Denmark and Norway (Kjørholt, 2005). These texts, published by the Danish Ministry of Culture, were of particular interest because of their rhetorical form, highlighting certain representations of children, freedom and self-realisation in discourses on children's participation (Kjørholt, 2005). I interpreted them as a public narrative about children's rights to decide for themselves and to realise themselves in 'free activities' with other children. Citizenship is represented as an individual's right to be free and make her or his own decisions. Freedom for the individual child is connected with notions of 'free choice', a core issue in the new practices that the staff members are implementing. Earlier daily routines, such as a fixed meal for everybody at a particular time, is one of the practices that are seen as being forced on children by adults, and therefore it is abolished. The overall argument is that the abolition of rules and of the structures of time and activities decided by adults creates a better life for children. It is further argued that listening to fairy tales by the whole group undermines children's rights to make their own decisions. Another example of the new practices that are implemented in the day-care centre is that toddlers are given 'the right to decide' when to change their nappies (Kjørholt, 2005). The construction of the autonomous 'toddler-citizen' in the Danish day-care centre provides something of a contrast to Mary's emotional testimony at the end of her performance and the narratives of the children in the 'Try Yourself' project. Mary was given the individual right to participate and to influence her life by speaking out against child-labour, but through her reaction she showed how this particular claim for autonomy undermined her achieving a better life for herself. As I have argued elsewhere, the toddlers and older children in the day-care institution are constructed in accordance with values that stress independence, individual choice and self-determination as ultimate goals (Kjørholt, 2005).

The particular way in which children are constituted as rights claimers from an early age in the two texts is, I suggest, an illustration of how global discourses on children's rights as social participants in this context are connected to particular moral values that are not explicitly discussed as such in the discourse. These overall values seem to take the form of the ability of children to make their own individual choices. In this particular narrative of childhood, time is a structuring element dividing the story of childhood into two phases: 'the past', characterised by a patriarchy controlling children's well-being in a negative way; and 'the present', including also future visions of equality for all, including children. The new practices that are being implemented are seen as an inevitable step in development, progressing towards democracy for all human beings. The authors take the view that the practices of the past must be left behind because they are oppressive to children and deny them their rights (Kjørholt, 2005). As such, the narrative is also about the development of egalitarian democratic societies, since it sees development as a kind of neutral and encouraging 'force' that is treated as politically and ideologically neutral, something to which one has to adapt.

The philosopher Charles Taylor has shown how in contemporary notions the modern subject is constructed as an individualised subject whose ultimate goal is self-realisation, and how this is a moral *ideal*. He argues that new moral positions are promoting the view today that everybody has the right to have their own values and to develop their own ways of life grounded in individual choices of what is important. This particular construction of the participating child subject is related in some ways to the global rights discourses and the emphasis on children as social participants. Taylor relates the centring on individual self-fulfilment to a moral idea of being 'true to oneself'. This moral ideal can be described as a *culture of authenticity*, pointing to a better or 'higher' mode of life. This higher mode of life is reached by subjects who are true to the 'inner voice' they are constituted with. In order to act correctly, one must listen to one's own nature and feelings 'deep inside'. Taylor criticizes the fact that the force of subjectivism and the contemporary culture of authenticity are not openly discussed as a moral ideal, but explained in terms of 'recent changes in the mode of production, or new patterns of youth consumption, or the security of affluence' (Taylor, 1991, p. 21).

The search for children's authentic voices both in the international society, as the global march and the voice of Mary Benjamin Olirata illustrates, and in national contexts, as the two presented participatory projects show, can be understood as part of realising contemporary moral ideals developed within the western world, of autonomous, authentic subjects as outlined by Charles Taylor. The particular construction of the participating child subject in the 'Try Yourself' project has clear commonalities with the construction of children as citizens in texts related to the Danish project entitled 'Children as Citizens'. In both projects, children were constructed as competent autonomous actors belonging to a specific age group and in need of being freed from adult control and power.

Self-determination was seen as a core issue and an overarching aim for children in day-care centres, as well as for their participation in local communities. In both projects, the search for children's 'authentic voices' was striking.

## Renewal of National Identity

Children are powerful symbolic actors, serving an increased demand for authentic voices in contemporary societies. In late modernity, children are envisioned as a form of 'nostalgia', not as 'futurity' (Jenks, 1996). Chris Jenks argues:

> Whereas children used to cling to us, through modernity, for guidance into their/our futures, now we through late-modernity, cling to them for 'nostalgic' groundings because such change is both intolerable and disorienting for us. They are lover, spouse, friend, workmate and, at a different level, symbolic representations of society itself (Jenks, 1996, p. 108).

With the 'linguistic turn' in social and human sciences, the significance of narrativity in the construction of individual identities, policy and cultural identities has been underlined. Narratives and texts have become increasingly important as constitutive of places, seen as physical locations that are socially constructed. This increasing importance not only concerns places, it is also of great relevance for other social phenomena, such as childhood, nation and democracy. Texts, images and symbols are therefore of vital importance in the construction of children as social participants or citizens.

The two participatory projects described above were followed by a huge marketing campaign in the local and national newspapers, television and the like. In Norway, a particular 'Try Yourself' information package was created, consisting of buttons, posters and leaflets delivered to school children with the aim of getting them to participate in the project. The huge advertising of 'Try Yourself', as well as children's desires to be pictured in the newspaper and to be visible in public spaces, underscores this growing significance of narrativity. Local newspapers played an important role in marketing the 'Try Yourself' project in general, as well as in writing about the various projects carried out by children in their local communities during the project period. The variety of different creative child-cultural activities initiated within the project represented excellent material for narrative constructions of children as creative, autonomous, competent actors in various public spaces, such as in different media. Constructions of local communities, national identity and traditional cultural notions of childhood were thus reconstructed and renewed.

My studies reveal that discourses on children as social participants in Norwegian society can be characterised as being in a position of hegemony, in the sense that they reflect certain constructions of the child that go relatively unchallenged and seem to be taken for granted (see Neumann, 2001). One condition for the existence of this hegemonic position is the link with international rights discourses that have become hegemonic, connecting the construction of the child subject to value concepts such as rights, freedom, choice, independence and individuality (Gullestad, 2003). These constructions of children as participants in society can be described as quite effective and almost impossible to resist in modern democratic societies, because the subject that is produced in this way is linked with values such as liberty, human rights, social equality, democracy, development and progress, all of which represent core values anchored in long traditions in western liberal societies. In many ways, this project also reflects core values in Norwegian society, not only concerning notions of a 'good childhood', but also of 'Norwegianness' and the national identity of Norway as a democratic nation. The reconstruction of this particular understanding of authentic childhood also serves to revitalise and reproduce the identity of Norway as a modern democratic nation that includes 'different' groups within a participatory democracy (Kjørholt, 2002, 2003). Children are therefore important actors in the renewal of Norway as a modern democracy. Their importance as vital actors contributing to the construction of 'imagined communities' in global societies in which traditional national and local identities and borders have become blurred is striking.

## Childhood as a Nation?

It has been argued that the concept of the nation is changing in late modern societies, and that new forms of 'nationhood' are being constructed as part of individual identities. Nationhood has been described as part of the 'inner self', one linked to individual identities. As a result of increased migration, boundaries between nations are being seen as blurred, and nationhood is no longer linked only to fixed physical boundaries. Instead it has become flexible and dispersed, as illustrated by the concept of the 'travelling nation' (Hultquist, 2001). Based on the powerful discourses on children as citizens in both the international society and national contexts, we may ask whether children are being constructed as a new 'global nation'.

In the project 'Try Yourself', children were constructed as an 'imagined community' supposed to inhabit 'their own authentic culture'. This 'imagined community of children' is, however, not anchored in any particular geographical area, but it is a notion of childhood constructed within particular historical and cultural circumstances. In 'Try Yourself', there were many applications from girls who wanted to engage in charitable activities of various sorts. The quotation below, taken from an application from four girls in the project, illustrates this point:

> We want to sing and make music cassettes, and if we are allowed to, we want to travel around, and that is going to be to the benefit of the cancer association (Elisabeth, Linn, Trude and Bjørg, 9–11 years).

In this text, the girls constitute an imagined community of friends, united in a shared interest in singing and making music, and in raising money for a charitable organisation, in this case a cancer association. Their construction of a *visible place* is not rooted in any specific geographical place. The words 'we want to travel around' clearly show that their place is spatially dispersed in the public space, like an imagined 'travelling community' of peers, and it vividly depicts the blurred boundaries between the local and the national. However, this travelling community not only represents a particular community of children, but also the locality the girls live in as a vital local community. Moreover, the text portrays the local community as a particular symbolic space for close social relations, charity and care. We may also ask whether contemporary discourses on children as active social participants are constructing new forms of nationhood in the global society.

## Children as Social Participants in Society: Hegemonic Constructions and Marginalised Positions

One of the conditions for the existence of the hegemonic character of discourses on children as participants in the Norwegian context is the inter-discursive relationship between discourses on democracy and nationality on the one hand, and children and childhood on the other. The terms 'child' and 'participation' both represent nodal points in the discourse, floating signifiers which different discourses fight to cover with meaning (Laclau and Mouffe, 1985). This implies that both terms can be filled with a variety of different meanings from other discourses occupying the field, yet still function as if the signifiers (the terms 'child' and 'participation') referred to one coherent explicit meaning. One important point is that this fact is often hidden in the discourse. The meanings of concepts such as children as 'active participants' and 'competent autonomous actors', used so often without further clarification or discussion, are, as I have demonstrated, often taken for granted. This taken-for-grantedness is further demonstrated by the overwhelming collective acceptance that these concepts seem to attain in different contexts, as well as the striking lack of discussion concerning their use.

There also seems to be a high degree of correspondence between the construction of the citizen in Norwegian discourses on democracy, and 'Norwegianness' as characterized by egalitarian individualism on the one hand (Bergreen, 1993; Eriksen, 1993; Gullestad, 1997) and the child subject in global human rights discourses on the other. The emphases on equality for all and the individual's rights to participate in society in Norwegian democracy are in line with the main principles embedded in the UN Convention on the Rights of the Child. The UNCRC therefore seems to function as a tool for the further development and strengthening of Norwegian democratic traditions (Kjørholt, 2002). The flourishing interest in discourses on children as social participants and citizens in Norwegian society, especially during the last 10–15 years, can largely be explained with reference to this inter-discursive relationship, which has contributed to making the discourse particularly powerful in producing truths that are taken for granted. According to Laclau and Mouffe, ideology plays a crucial role in the construction of hegemony (Torfing, 1999, p. 113).

Despite being placed in hegemonic discourses on children as social participants in society in recent years, children are still being situated in marginal positions in many ways. The great majority of the many participatory activities and projects initiated in Norwegian municipalities within a 10-year period were *ad hoc* and short term, and they suffered from a lack of integration into permanent and legal structures. The fact that more than 60% of these participatory activities were addressed to children aged 14 or more reveals that age is still a category producing exclusion from different kinds of participation in society, and even one's marginalisation (Kjørholt, 2004).

The results of a survey of participatory projects in Norway also reveals other interests than children's rights to participation as being involved in discursive practices on children and participation (Kjørholt, 2002). Important aims associated with the construction of children as competent participants in society were the creation of drug-free environments, the construction of sustainable local communities by preventing young people from moving, and the protection of the environment. These results illustrate the arguments presented by Laclau and Mouffe (1985), that nodal points are 'floating signifiers' which other discourses fight to cover with meaning. This vulnerability of discourses on children as social participants in society easily places children in marginal positions.

## Childhood as a Symbolic Space: Concluding Discussions

In this chapter, I have discussed participatory projects relating to global discourses on children's rights to participation and citizenship in Nordic countries in recent years. I have presented glimpses into two different participatory projects, arguing that discourses on children as social participants in society constitute childhood as a symbolic space, representing and reproducing symbolic values related to democracy, national identity, autonomy and authenticity. I have further argued that the growing accent on children as social participants and citizens in Norway since the 1990s is connected to the dynamic interplay between international children's rights discourses and particular cultural constructions of children and childhood (especially ideas of what constitutes a 'good' childhood) operating on the national as well as local levels in the Norwegian context. The emphasis on children as competent social participants, revealing their agencies and their worth of being studied in 'their own right', is pertinent in making visible their contribution to the economic, political and cultural (re)production of society. The growing interest in children's experiences and perspectives within both research and policy fill a gap of knowledge regarding the reality of a group of human beings that has too long been neglected. However, the notion of 'children's own culture', seen as a distinct culture separated from the surrounding adult cultural context, may contribute to concealing the dynamic interrelatedness between children's meaning-making processes and social practices as these are developed among peers in different places, and the broader cultural and political structure, mainly constituted by adults, in which children are embedded. The increasing emphasis on 'giving a voice' to a 'different other' easily turns into essentialistic constructions of children, neglecting the dynamic new ways in which children and childhood are embedded in economic, political and socio-cultural transformations on a macro-scale. Moreover, the increasingly symbolic value of children and childhood in the era of global change may thus be hidden. This perspective is also, I suggest, highly pertinent in the analysis of 'children's geographies'. Children's agency and social practices in different places on the micro-scale are influenced by and constituted within a web of complex interrelationships with adult's social practices and places. Childhood is to an increasing degree constituted as a symbolic space representing particular moral values in an era of extensive economic, social and political change. There is a danger of these values and this symbolic 'nature' being overlooked if children's geographies, experiences and voices are studied on the micro-scale without including the broader cultural context within the analysis.

The narrative of children's rights to participation within Danish day-care institutions is a narrative of particular forms of individualism in a modern Western society that constructs certain ideas of individual autonomy and self-realisation as overarching moral values. Self-realisation is conceptualised as the individual's right to make her/his own choices and decisions. Children's self-realisation is first and foremost seen as an individual project that can be realised within an age-related social order. As such, the narrative I have entitled 'The right to be oneself' is also a public narrative that conceptualises an age-related social order as a moral ideal, constructed as a relationship between equal and individual child subjects. Play is seen as a core activity of subjects belonging to a 'community of children', reflecting particular cultural notions of what it means to be a child in the Danish context. This particular construction of children and childhood is in line with constructions of children as social participants and fellow citizens in Norway (Kjørholt, 2001). Olwig and Gulløv suggest that the description of children and adults as separate so often found in the research literature is not a universal characteristic but one anchored in western studies and notions of childhood (Olwig and Gulløv, 2003). They argue: 'Children are not necessarily marked as a distinct group defined in contrast to adults, and we therefore need to examine closely the nature of relationships between people of varying ages in different cultural settings' (2003, p. 13).

The project 'Try Yourself' also demonstrated that children are active social participants in reproducing national identity and notions of Norway as a democratic nation, a fact that went far beyond the scope of the project. More than that, my main argument is that children today are also significant social actors contributing to the reproduction of notions of national identity, democracy and authenticity in a world where nationality, culture and boundaries between childhood and adulthood as two distinct and separate life worlds are being contested. The voice of Mary, a child labourer and child activist, clearly demonstrated paradoxes and challenges related to the promotion of children as autonomous social participants within rights discourses. Contemporary childhood is to an increasing degree constituted as a symbolic space, contributing also to reproducing notions of human beings as authentic, autonomous subjects. Childhood then operates as a targeted space for the reproduction of symbolic values relating to participatory democracy, nationality and authenticity.

The symbolic importance of children as participants is increasing in modern societies, and their symbolic value has been underestimated. This increasing symbolic value of children as social participants makes it highly important to discuss how these new discourses affect children's livelihoods in different parts of the world. Furthermore, the hegemonic character of global discourses on children's rights to participation, tied up with the increasingly symbolic character of children and childhood, makes it highly pertinent to include analysis of the ideology and moral values that these discourses represent. So far, such critical analysis of global rights discourses has largely been lacking.

Much of the literature on children's participatory rights is characterised by universalising and normative assumptions (Hart and Chawla, 1982; Langsted, 1992; Poulsgaard, 1993; Verhellen, 1993; Flekkøy, 1993; Franklin, 1994; Pavlovic, 1994; Van Gils, 1994; Alderson, 1999) about the self-evident value of children's participation, rather than providing a critical scrutiny of political discourses on the implementation of particular projects, or focusing on the actual experiences of child participants in these projects. International comparative studies of the topic are rare (Horelli, 1998), and a consensus on common terminology and theory at the international level is badly lacking (Riepl and Wintersberger, 1999). These studies urge further empirical investigation of how participation rights and participatory projects affect children's lives, and how universal rights are implemented in different social and cultural contexts. This is in addition to further calls for research to develop theoretical and conceptual clarifications of notions of citizenship and rights to participation.

## Notes

1. In both Denmark and Norway, most children aged 1–6 are placed in institutional care such as day-care centres (Norway: *Barnehage* 1–6 year-olds: 66%). (SSB 2002), Denmark: *Børnehaver*: 90%, *vuggestue*, 1–6-year-olds: 60%) The children spend approximately six to nine hours a day in the institution. In both Denmark and Norway, the curriculum in day-care institutions emphasises 'free play' to a large extent, underlining the fact that the pedagogy is more child-centred than in primary schools.
2. This concept is often used in different ways without further clarification. Elisabeth Näsman, referring to Turner 1986, distinguishes between three forms of individualism: firstly a political doctrine of individual rights; secondly an expression of individual autonomy; and thirdly the process of individuation, which points to integrative processes connecting the individual to social forms (Näsman, 1994). It is the first two forms that are of particular relevance for my discussion here.

## References

Bråten, S. (1998) *Intersubjective Communication and Emotion in Early Ontogeny*, Cambridge: Cambridge University Press.

Enerstvedt, Å. (1971) *Kongen over gata*, Oslo: Universitetsforlaget.

Flekkøy, M.G. (1993) *Children's Rights: Reflections on and Consequences of the use of Developmental Psychology in Working for the Interest of Children. The Norwegian Ombudsman for Children: A Practical Experience*, Ghent: Ghent University.

Franklin, B. (1994) Children's rights to participate. Paper presented to the European Conference on Monitoring Children's Rights. Ghent, 11–14 December 1994.

Gullestad, M. (1997) From being of use to being oneself: dilemmas of value transmission between the generations in Norway, in M. Gullestad and M. Segalen (eds) *Family and Kinship in Europe*, London: Pinter, 202–18.

Gullestad, M. (2003) Fighting for a sustainable self-image: the role of descent in individualized identification, *Focaal: European Journal of Anthropology*, 42, 51–62.

Gulløv, E. (2001) Placing children. Paper presented to a research seminar on 'Children, Generation and Place: Cross-cultural Approaches to an Anthropology of Children', 19–21 May 2001, University of Copenhagen, Network for Cross-Cultural Child Research.

Hart, R. (1992) Children's Participation: From Tokenism to Citizenship. United Nations Children's Fund. *Innocent Essays*, 4.

Haugen, S.P. (1995) *Barn som medborgere: Visjon og virkelighet*, Hovedoppgave. Universitetet i Bergen: Institutt for kunsthistorie og kulturvitenskap, avd. for etnologi og folkloristikk.

Hengst, H. (2006) *In Flexible Childhoods?* University Press of Southern Denmark, Odense.

Horelli, L. (1998) Creating child-friendly environments: case studies on children's participation in three European countries, *Childhood*, 5(2), 225–39.

Hultquist, K. (2001) *Governing the Child in the New Millenium*, London: Routledge.

Hviid, P. (1998) Deltakelse eller reaktiv pædagogik, in U. Brinkkjær *et al.* (eds) *Pedagogisk faglighed i dagistitusjoner*. Rapport 34, København: Danmarks pædagogiske universitet, 207–26.

Højlund, S. (2000) Childhood as a social space: positions of children in different institutional contexts. Paper presented to the conference, 'From Development to Open-ended Processes of Change', 6–7 April 2000, University of Copenhagen, Institute of Anthropology.

James, Jenks and Prout (1998) *Theorising Childhood*, Cambridge: Polity Press.

Jenks, C. (1996) *Childhood*, London: Routledge.

Junker, B. (1998) *Når barndom bliver kultur*, København: Forlaget Forum.

Kampmann, J. (2001) Hva er børnekultur? in B. Tufte, J. Kampmann and B. Junker (eds) *Børnekultur. Hvilke børn og hvis kultur?* København:Akademisk Forlag A/S.

Kampmann, J. (2003) Udviklingen af et børnekulturelt pædagogisk blikk, in Birgitte Tufte, Jan Kampmann and Marit Hassel (eds) *Børnekultur: et begreb i bevægelse*, København: Akademisk forlag, 86–98.

Katz, C. (2004) *Growing up Global: Economic Restructuring and Children's Everyday Lives*, Minneapolis, MN: The University of Minnesota Press.

Kjørholt, A.T. (2001) 'The participating child': a vital pillar in this century? *Nordisk Pedagogikk—Nordic Educational Research*, 21(2), 65–81.

Kjørholt, A.T. (2002) Small is powerful: discourses on 'children and participation' in Norway. *Childhood 1, Volume 9*, 63–83.

Kjørholt, A.T. (2004) Childhood as a social and symbolic space: discourses on children as social participants in society. Ph.D Thesis. Trondheim: NTNU.

Kjørholt, A.T. (2005) The competent child and the 'right to be oneself': reflections on children as fellow citizens in a day-care centre, in A. Clark, A.T. Kjørholt and P. Moss (eds) *Beyond Listening: Children's Perspectives on Early Childhood Services*, University of Bristol: The Policy Press.

Kjørholt, A.T. and H. Lidén (2004) In H. Brembeck, B. Johansson and J. Kampmann (eds) *Beyond the Competent Child: Exploring Contemporary Childhoods in the Nordic Welfare Societies*, Roskilde: Roskilde University Press.

Laclau, E. and C. Mouffe (1985) *Hegemony and Socialist Strategy: Towards a Radical Democratic Politics*, London and New York: Verso.

Langsted, O. (1992) Valuing quality: from the child's perspective, in P. Moss and A. Pence (eds) *Valuing Quality*, København: Hans Reitzels forlag, 28–42.

Mortier, F. (2002) The meaning of individualization for children's citizenship, in F. Mouritsen and J. Qvortrup (eds) *Childhood and Children's Culture*, Odense: University Press of Southern Denmark, 79–102.

Näsman, E. (1994) Individualization and institutionalization of childhood in today's Europe, in J. Qvortrup, M. Bardy and H. Wintersberger (eds) *Childhood Matters: Social Theory, Practice and Politics*, European Centre Vienna 14. Aldershot: Avebury, 165–88.

Neumann, I. (2000) *Mening, materialitet, makt: en innføring i diskursanalyse*, Oslo: Fagbokforlaget.

Pavlovic, J. (1994) The Children's Parliament in Slovenia. Paper Presented to the European Conference on Monitoring Children's Rights. Ghent, 11–14 December 1994.

Poulsgaard, K. (1993) Das Projekt Kinder als Mitburger, in H. Hengst (ed.) *Von, für und mit Kids: Kinderkultur in europaischer Perspektive*, Munchen: Kopad Verlag.

Riepl, B. and Wintersberger, H. (1999) Towards a typology of political participation of young people, in B. Riepl and H. Wintersberger (eds) *Political Participation of Youth Below Voting Age*. Eurosocial Report 66. Vienna: European Centre for Social Welfare Policy and Research, 225–38.

Prout, A. (2000) Children's participation: control and self-realisation in British Late Modernity, *Children and Society*, 14(4), 304–15.

Ried Larsen, H. and Larsen, M. (1992) *Lyt til Børn: en bok om børn som medborgere*, København: Det tværministerielle Børneudvalg og Kulturministeriets Arbejdsgruppe om Børn og Kultur.

Shore, C. and Wright, S. (1997) Policy: a new field of anthropology, in C. Shore and S. Wright (eds) *Anthropology of Policy: Critical Perspectives on Governance and Power*, London: Routledge, 3–42.

Selmer-Olsen (1990) *Barn imellom og de voksne: en bok om barns egen kultur*, Oslo: Gyldendal.

Stephens, S. (1995) Children and the politics of culture in 'Late Capitalism', in S. Stephens (ed.) *Children and the Politics of Culture*, Princeton, NJ: Princeton University Press, 3–44.

Tatek A. (2007) Changing livelihoods, changing childhoods: patterns of children's work in rural southern Ethiopia, *Children's Geographies*, 5(1–2), 77–93.

Taylor, C. (1978) *Sources of the Self: The Making of the Modern Identity*, Cambridge: Cambridge University Press.

Taylor, C. (1985) *Philosophy and the Human Sciences*, Philosophical Papers 2. Cambridge: Cambridge University Press.

Taylor, C. (1991) *The Ethics of Authenticity*, Cambridge, MA, and London: Harvard University Press.

Threvarthen, C. (1973) The function of emotions in early infant communication and development, in J. Nadel and I. Camanioni (eds) *New Perspectives in Early Communication and Development*, London: Routledge.

Tingstad, V. (2003) Children's chat on the net. Ph.D Thesis, Trondheim: NTNU.

Tingstad, V. (2006) *Barndom under lupen: Å vokse opp i en foranderlig mediekultur*, Oslo: Cappelen Akademisk forlag.

Torfing, J. (1999) *New Theories of Discourse: Laclau, Mouffe and Zizek*, Oxford: Blackwell.

Van Gils, J. (1994) Snater: une experiénce d'autogestion par les enfants quelle est sa valeur mediate. Paper presented at the European Conference on Monitoring Children's Rights. Ghent 11–14 December 1994.

Verhellen, E. (1993) Children and participation rights, in P.L. Heiliö, E. Lauronen and M. Bardy (eds) *Politics of Childhood and Children at Risk: Provision, Protection and Participation*, Eurosocial Report 45. Vienna: European Centre for Social Welfare Policy and Research, 49–64.

Woodhead, M. (1999) Reconstructing developmental psychology: Some first steps, *Children and Society*, 13(1), 3–19.

# Childhood in the Age of Global Media

DAVID BUCKINGHAM
*Institute of Education, University of London, London, UK, and Norwegian Centre for Child Research, NTNU, Trondheim, Norway*

In most regions of the world, the media are now an inescapable fact of contemporary childhoods. In most industrialised countries, children spend more time with media of various kinds than they spend in school, or with their family or friends. Even in the rural areas of developing countries, the advent of electronic media is often an early harbinger of 'modernisation'; and growing numbers of children have access to globally- and locally-produced media material. However, the role of the media—and, more broadly, of children's consumer culture—has typically been neglected by sociologists of childhood: definitive overviews of the field often make little or no mention of the issue (e.g., James *et al.*, 1998). As Sonia Livingstone (1998, p. 438) has argued, the new sociological child appears to live a 'natural', *non-mediated* childhood: this is 'a care-free child playing hopscotch with friends in a nearby park, not a child with music on the headphones watching television in her bedroom'.

Historically, research on children's relationships with media has been dominated by psychological perspectives. Broadly behaviourist studies of the media's effects on behaviour or attitudes—most notably in relation to 'violence'—have partly given way to a more constructivist approach, in which the child is seen as an active processor of meaning, rather than a passive victim (Singer and Singer, 2002; van Evra, 2004). In general, however, researchers in this field continue to employ a developmentalist approach, in which children are seen to be gradually progressing towards a state of adult rationality. The focus here is on the interaction between the individual child and the screen, in isolation from broader interpersonal and social processes.

In recent years, however, researchers in the fields of Media and Cultural Studies have been developing a more sociological account of children's uses and interpretations of media (see, for example, Gillespie, 1995; Buckingham, 2000; Livingstone, 2002). This research seeks to explore children's perspectives, and to analyse their interactions with media, on their own terms. In the process, it draws attention to children's *competence* as media users—their 'media literacy'—and the ways in which media use is embedded in the contexts and relationships of everyday life. However, it also recognises that children's dealings with media are framed by the operations of the media industries, and by the constraints exerted by textual meanings: children are by no means simply free to make their own meanings in any way they choose. Understanding the power of the media thus requires us to theorise and account for the complex relationship between structure and agency in defining childhood (Buckingham and Sefton-Green, 2004).

In this essay, I intend to apply this kind of approach to analysing the place of the media in children's experiences of globalisation. As I shall argue, children are not merely passive victims of all-powerful media representations; but neither are they completely free agents. Both economically and culturally, the media play a profoundly ambiguous role in terms of globalisation: they provide powerful and pleasurable forms of 'children's culture' that appeal to children living in very different circumstances around the world; yet they also provide symbolic resources with which children come to define their own meanings and identities.

## Media and Globalisation

Like the trade in material goods, the trade in cultural goods is undoubtedly a key factor in the contemporary reconfiguration of relations between the global and the local. Nevertheless, there are some starkly opposing accounts of this process, which beg broader questions about our understanding of the power of the media and of the ways in which it is exercised.

Thus, on the one hand, we have theories of *cultural imperialism*, which point the finger of blame directly at the United States, as the world's leading superpower. As the title of one influential early book expressed it, 'The Media Are American' (Tunstall, 1977). From this perspective, US media are powerful agents of cultural homogenisation: they eradicate local or indigenous cultures by imposing a singular ideology and world-view. This development is seen as an inevitable consequence of capitalist expansion, as corporations restlessly seek out new markets, and as economies of scale result in a steady growth of monopolisation. Rather than relying simply on physical occupation, the US is now seen to sustain its hegemony through a process of ideological and cultural domination, or 'Coca-colonisation' (Wagnleitner, 1994).

This kind of argument has been widely criticised by many scholars. It is argued that the flow of cultural goods is not so straightforwardly unidirectional; and that *economic* power does not necessarily result in a form of *ideological* domination. Tomlinson (1991), for example, argues that such arguments effectively infantilise consumers, implying that they are somehow powerless to resist colonial ideologies; and he points to evidence from audience research that shows the diverse ways in which global audiences respond to (and, in many situations, resist) US-made cultural products (e.g., Liebes and Katz, 1990). Researchers also point to the dominance of home-grown media products in domestic markets (Silj, 1988); the historical and continuing popularity of non-US cultural products around the world (French paintings, Chinese food, Latin American music) (Cowen, 2002); and the emergence of new 'cosmopolitan' global cultures (Hannerz, 1996).

In more recent years, the cultural imperialism thesis has effectively given way to a much more optimistic account of the global spread of media. Rather than the global replacing the local, the two have been seen to merge in a process of 'glocalisation' (Robertson, 1994; Featherstone, 1995); and, it is argued, this process also has a very long history, rather than being a development unique to contemporary capitalism. Advocates of this approach point to the flourishing of local cultural production in many regions of the world; and to the global dissemination of non-US media products, ranging from Brazilian *telenovelas* to Japanese animation to Jamaican reggae music (Cowen, 2002). They also proclaim the new forms of 'hybridity'—or 'new ethnicities'—that emerge as global media forms (such as hip-hop: Bennett, 2000) are merged with local idioms and traditions. From this perspective, globalisation actively produces cultural diversity, rather than homogeneity; and cultural identities are accordingly fluid and open to change.

However, critics have begun to point to some of the difficulties with this more optimistic approach. They question the emphasis on 'hybridity', arguing that this is not equally available to all, and that it may be merely a characteristic of the intelligentsia rather than of the population as a whole (Cohen, 1997); they critique the commodification that continues to characterise the global trade in culture, for instance in the marketing of so-called 'world music' (Feld and Keil, 1992); and they challenge the superficiality (or 'ethnic chic') of some such developments (Leshkowich and Jones, 2003). It is certainly possible to argue that this more optimistic account neglects the economic dimension of the media—that is, their function as a means of generating profit for already-wealthy nations, which was a key concern of the cultural imperialism thesis—and slides into a somewhat easy form of celebration that is characteristic of some postmodern cultural theory.

## Global Childhoods

What happens when children enter this picture? The global scale of marketing to children typically provokes an additional anxiety, which is essentially about cultural continuity. According to the critics, what we are seeing is the construction of a homogenised global children's culture, in which cultural differences are being flattened out and erased, and in which parents' attempts to sustain their cultural values are increasingly in vain. The media are seen to have disrupted the process of socialisation, upsetting the smooth transmission of values from one generation to the next. For authors such as McChesney (2002) and Kline (1993), this is an inevitable consequence of commercialisation: neo-liberal economics are, in their view, inherently incompatible with the 'real' needs and interests of children. Within the media production sector, there are also frequent calls for government intervention to support indigenous children's production against the pressure of the global market (von Feilitzen and Carlsson, 2002).

Such arguments almost inevitably have an element of conservatism, and often seem to rely on judgments about cultural value that are asserted rather than fully justified (see Buckingham, 2000, Chapter 7). By contrast, it could be argued that the media are responsible for a *modernisation* of childhood, or at least for the growing dominance of a modern*ist*

discourse about childhood. For example, the global success of the children's channel
Nickelodeon provides a symptomatic instance of the ways in which market values have
come to be aligned with liberal political arguments about children's rights, of the kind
described by Tracey Skelton and Anne Trine Kjorholt elsewhere in this issue. The statements of Nickelodeon executives and the rhetoric of its on-screen publicity proclaim its
role as an agent of empowerment—a notion of the channel as a 'kid-only' zone, giving
voice to kids, taking the kids' point of view, as the friend of kids; and the interests of
'kids' are frequently defined here as being in opposition to the interests of adults (see
Hendershot, 2004).

This 'modernist' discourse has increasingly been adopted by exponents of more traditional children's media, not least in public service television. This is certainly the case
in the UK (see Buckingham *et al.*, 1999); but it is also apparent, for example, in the (possibly unlikely) case of the Minimax channel in post-communist Hungary (Lustyik, 2003). It
might even be possible to talk here about the emergence of a globalised 'modernist style' in
children's culture, that seeks to address children across cultural boundaries—in the same
way that critics have talked about 'youth culture' as a global phenomenon (Buckingham,
2004). The Japanese business consultant Kenichi Ohmae (1995), for example, asserts that
children's exposure to such a global culture means that they now have more in common
with their peers in other cultures than they do with their own parents.

Of course, we should not forget that many children in the majority world—particularly
in the more isolated rural areas of poorer countries—still do not have access to many forms
of media communication. Such children and their families may lack the economic
resources to purchase manufactured toys or printed media, and may lack electricity,
let alone access to broadcast signals; and as such, their ability to 'buy into' a globalised
children's culture is distinctly limited (for example, see Punch, 2003, on rural Bolivia).
Yet as modern technologies steadily extend their reach across the globe, the media that
these children are likely to encounter first will not be those produced in their own
countries, but those produced in the wealthy nations of the West.

## The Economic Case

There is certainly an economic logic to the globalisation of children's culture. The children's market is potentially large, but it is also by its nature quite fragmented. In terms
of age, children are quite clearly divided (and divide themselves) into age segments.
What appeals to a five-year-old is unlikely to appeal to a 10-year-old.[1] The market is
also clearly divided in terms of gender. Particularly for younger children, this is a very
'pink and blue' market, and there are significant risks in attempting to cross the line, in
order to appeal to both groups. It used to be the received wisdom among marketers that
the way to succeed was to appeal to boys first: girls were quite likely to buy into boy
culture, although boys were less likely to buy into girl culture (Schneider, 1992). Analyses
of contemporary toy advertising would suggest that—despite decades of second-wave
feminism—this continues to be the case (Griffiths, 2002).

While the picture is (perhaps increasingly) complicated, the fact remains that the children's market is strongly segmented; and one way for producers to deal with this—to
develop economies of scale—is to build markets globally, to amass 'niche' markets
into a larger global market. This tendency is also reinforced by the vertical and horizontal
integration of the media and cultural industries. 'Vertical integration' refers to the way in
which the market is coming to be dominated by a small number of global players, who
integrate hardware, software and means of distribution; and these are largely the same
players who dominate the adult market (Westcott, 2002). For example, in the case of

specialist children's TV channels, the market is dominated by Disney (who have significant interests in the adult market via subsidiaries like Touchstone and Buena Vista, and own the ABC network in the US), Nickelodeon (owned by Viacom), Cartoon Network (owned by AOL Time Warner), and Fox Kids (Murdoch). It is these US-based companies that dominate the children's market: these four companies run more than thirty branded children's channels across Europe, for example, although none of them invests to any significant degree in local production. This is by no means a risk-free market, and local producers are responding to the challenge; but in homes where children have access to cable/satellite, these US-owned channels are achieving a growing market share.

These companies are also increasingly operating across media platforms ('horizontal integration'). Nearly all the major children's 'crazes' of the last twenty years (Ninja Turtles, Power Rangers, Pokémon, Beyblades, Harry Potter, Yugioh) have worked on the principle of what the industry calls 'integrated marketing', or multimedia 'synergy'. For example, Pokémon was first a computer game, then a TV series, a trading card game, a series of movies, and a whole range of merchandise, from clothes and toys to food and bags and all sorts of unlikely paraphernalia. Again, horizontal integration on this scale requires global marketing: it would be much harder to achieve in one country alone.

As this implies, there is an economic logic to the globalisation of the children's market; and to this extent, the cultural imperialism thesis appears to be correct. Yet *economic* domination does not necessarily translate directly into *ideological* or *cultural* domination. Sam Punch's article in this issue shows how, as individuals move across cultures, the purchase of certain commodities (such as clothes) may accompany changes in attitudes; but even here, the direction of any causal relationship between these things is not necessarily easy to establish. In the cultural sphere, globalisation is often a paradoxical phenomenon. As I intend to show, children's culture is characterised not so much by a one-way process of domination, but by an unpredictable and contested relationship between the global and the local—often expressed in the notion of 'glocalisation' (Robertson, 1994). In the following sections of this article, I explore this further by focussing on some specific examples: I begin with two instances of commercial culture, Disney and Pokémon, and conclude by looking briefly at a couple of 'public service' productions.

## Uncle Walt and His Evil Empire

For critics of commercialism, Disney is the 'bad brand' of children's culture. It is the MacDonalds, the Nike of childhood. Historically, it is hard to ignore the political intentions of the Disney corporation. Uncle Walt's involvement in far-right politics is quite well documented (Roth, 1996); and critics have seen some of the early films as little more than ideological propaganda in the service of US foreign policy. Eric Smoodin's (1994) collection, for example, contains some trenchant critiques of the role of Disney movies promoting cultural imperialism in Latin America: the Donald Duck film *The Three Caballeros* is singled out for particular criticism (Burton-Carvajal, 1994; Piedra, 1994), alongside some of the lesser-known travelogues and educational films (Cartwright and Goldfarb, 1994).

Donald Duck was also the subject of a famous early critique by Ariel Dorfman and Armand Mattelart (1975), whose book *Reading Donald Duck* was written in Chile in the wake of the United States' involvement in the overthrow of the Allende government. They argued that, far from being a harmless fantasy of childhood innocence, Donald was actually a vehicle for the propagation of capitalist ideology and US interests and values. That critique has certainly been questioned, however. Martin Barker's (1990) book on comics takes Dorfman and Mattelart to task for a kind of conspiracy theory that over-emphasises the closeness of the relationship between government and business; and, perhaps more importantly, for oversimplifying the ways in which people read these comics—in effect, for assuming that children would simply swallow whole the values that the comics were seen to represent.

Even so, for many adults—particularly middle-class adults—outside North America, Disney is synonymous with the unacceptable face of US capitalism. My own research on this was undertaken as part of a global project, published in the book *Dazzled by Disney?* (Wasko *et al.*, 2001). For the older adults (many of them parents) whom we inter-viewed, Disney was seen as safe, sanitized, predictable, inauthentic and somewhat 'corny' (Buckingham, 2001). It was full of objectionable stereotypes and cheap moralism; and it was leading children away from true (that is, national) cultural traditions towards a homo-geneous, mass-produced consumer culture. Disney was all about brainwashing, seducing the innocent—and in this sense, the debate about Disney was tied up with broader argu-ments about the changing symbolic value of childhood. This is a view that is also apparent in some of the academic criticism, for example in the work of Henry Giroux (2001)—although for our British respondents, it was very much reinforced by a sense of Disney as quintessentially *American*, as somehow inherently alien. Their resentment of Disney—as of Macdonalds—was often inextricably connected with their rejection of the United States' political role as a global superpower.

However, this criticism was not without its contradictions and paradoxes. Some of the younger adult participants we interviewed (who were mostly undergraduates) were more inclined to admit to enjoying Disney: they expressed a kind of aesthetic appreciation of the animation and the spectacle, and talked about the emotional appeal of Disney's construc-tion of childhood. Even for the older adults, there was a certain amount of ambivalence, even tinged with a degree of nostalgia. It may be that the symbolic importance of Disney as a bearer of US capitalist values tends to provoke a principled critical discourse, even where that discourse might not actually reflect how individuals respond to the films themselves.

Nevertheless, this sense of Disney's 'Americanness' was not at all an issue for the chil-dren we interviewed. When I asked one group of six year olds where they thought Disney films came from, they were quite unsure (despite the US accents). In the end, they opted for France, on the grounds that one of them had been to Disneyland Paris. A similar finding was apparent from Kirsten Drotner's (2001) research for the same project in Denmark. Her chapter in the book is named after a quote from one of her respondents: 'Donald seems so Danish'. To some extent, she is suggesting that the Americanness—the alienness—of Disney went unnoticed by the young Danes whom she interviewed, partly because the comics are translated and the movies are dubbed (which obviously is not the case in the UK). However, she also suggests that they read the comics selectively: the good aspects of Donald (his unconventional, politically incorrect style) were read as Danish, as reflect-ing some kind of national character, while his bad qualities were seen as American, and hence as 'other'.

On one level, of course, this invisibility could be seen precisely as testimony to the power of Disney's cultural imperialism—that, if the media are American, then we are all American too. It could be that Americanness is the default position, something so uni-versal and so unquestioned that it has become effectively invisible. On the other hand, it could be argued that, as they increasingly engage with world markets, Disney and other cultural producers are having to suppress elements that might be perceived to be too cul-turally specific in favour of those that seem to speak to some universal, trans-cultural notion of childhood.

There are some elements of truth in both arguments. In fact, what our research and Drotner's shows is that audiences read selectively, taking aspects of the text that seem to them to confirm a positive self-image or cultural identity, and setting that against other aspects of the text that they perceive as problematic or undesirable. And it is these latter aspects that they often seem to define as 'American'—as plastic, fake, kitschy, or indeed as ideologically suspect. Obviously, this is also an historical process: it would certainly be interesting now to map young people's changing perceptions of the US, as they are mediated through popular culture, in the wake of George Bush's so-called 'war on terror'.

However, the actual texts that are being produced for the global market are also more complex. Politically speaking, the key Disney movies of the last couple of decades have moved quite a long way from *The Three Caballeros*. McQuillan and Byrne (1999) read the Disney of the 1980s and 1990s as a set of parables for US foreign policy—so *The Lion King* is about South Africa, *Aladdin* is about the Middle East, *Mulan* is about China, and so on. On one level, it is possible to read all these texts as being about global ideological colonisation—and, if you add in *Pocahontas*, about internal colonisation as well. One might argue that they are all about recuperating and sanitising 'other' cultures and the challenges they might represent. However, I believe it makes more sense to see them as liberal texts, even as confused liberal texts—as texts which do at least recognise other cultures, and which try to construct some kind of dialogue with them. If we see such films merely as a more subtle form of propaganda (as critics like Giroux are wont to do), then we miss much of their ambivalence and complexity.

## Gotta Catch 'Em All

Pokémon provides an interesting contrast with Disney, on several levels. Walt Disney always maintained that his films were intended for a family audience, and not just for children; and while this was essentially a financial decision on Disney's part, it is possible to argue that the films tell children and adults rather different stories about childhood (Forgacs, 1992). By contrast, Pokémon was largely inaccessible, impenetrable even, to the majority of adults; and that in itself partly explains some of its appeal to children.

My own research on Pokémon was part of another global project, which involved researchers from Japan and the US as well as from the UK, France, Israel and Australia. This research is collected in Joseph Tobin's book *Pikachu's Global Adventure: The Rise and Fall of Pokémon* (Tobin, 2004). Among other things, we were interested in how Pokémon was produced and marketed as a global phenomenon, and how it was perceived and used by children. As I have noted, Pokémon began life as a computer game, which was produced by Nintendo, but it rapidly span off into a TV series, movies, books and comics, a trading card game, toys, plus the usual range of clothing, food, lunchboxes, stickers and countless other products. In many ways, it was a classic example of integrated marketing. The Pokémon brand became a means to market a whole range of licensed goods to a very diverse range of audiences: soft toys for younger children, TV cartoons for a slightly older age group, the Game Boy game for the oldest, pre-teens and younger teenagers. There was also appeal to girls as well as boys, with girl-friendly themes and characters as well as the more predictable boy-oriented ones: so Pokémon was about collecting, about nurturing as well as competing, about feeling as well as fighting. For a while at least, it managed very effectively to cut across the divisions and segments in the children's market.

It was also very successful globally. This needs to be understood, firstly, in terms of the emergence of Japan as a global cultural power—which has become an increasingly important factor in global media markets (see Allison, 2000). This phenomenon is most obviously apparent in the rise of *manga* (the Japanese comic book) and its move from being a 'cult' medium, or a medium just for children, to something more mainstream; and it is also evident in the global marketing of Japanese 'cuteness' (*kawai'sa*), in the form of Hello Kitty, Sailor Moon, and even the Tamagotchi (Yano, 2004). Pokémon clearly combines elements of both these forms.

Our research considered how Japanese producers used other countries as a kind of springboard into local markets around the world. In Asia, where there can be resistance to Japanese products, Nintendo marketed Pokémon products via Hong Kong; while it also used the United States, and US franchise holders, to push into other Western markets. So the Pokémon card game was actually produced by a US company, the sinisterly-named Wizards of the Coast, who were also responsible for a similar game for older children, *Magic the Gathering*. Likewise, the version of the TV series that is exported globally is a re-versioned one from the US, not the Japanese original: it is an 'Americanised' version of a Japanese cultural product.

This 'glocalisation' involved two things. Firstly, at the point of distribution, there is a process that the industry calls 'localising'. Thus, the distributors employed people in the US to adapt the TV cartoons for the US market, which involved editing out material that was seen to be too culturally strange or specific. Hirofumi Katsuno and Jeffrey Maret (2004) compared the Japanese and US versions of Pokémon cartoons and found some quite unexpected (or at least quite bizarre) differences here, in terms of what was seen to be palatable to a US market—particularly to do with the removal of anything remotely sexual or 'violent'.

At the same time, at the point of production, there was a process Koichi Iwabushi (2004) calls 'deodorising': that is, anything that was seen as too Japanese (the 'odour' of Japan) was removed. So, for example, the characters are given Westernised names, the characteristic *anime* visual style (for example, of the faces) is relatively muted, there is no written language, no religious references, and some of the settings are rendered less obviously Japanese. There was therefore a distinct effort to create a product that would be exportable, both at the point of local distribution and at the point of production.

To some extent, as Kenichi Ohmae (1995) suggests, this may facilitate the creation of a common global culture among children. We observed several instances of children communicating and playing with Pokémon across social and cultural differences: its extensive, specialised mythology provided a kind of common language that transcended cultural barriers. Yet what was interesting was that—unlike my brief experience with Disney—the 'Japaneseness' of Pokémon was something that children definitely recognised and enjoyed. It was strange, but strange-exotic—and, indeed, profoundly cool (see McGray, 2002). Of course, 'cool' has a limited shelf-life; and our project also tracked the complicated reasons why children appeared to drop Pokémon as quickly as they had taken it up. Even so, it would be interesting to consider how the continuing rise of Japan in the cultural sphere might play out in terms of children's global awareness.

## Glocalisation as Public Service

This phenomenon of 'glocalisation' is not entirely new, nor are the processes I am describing confined to the more obviously commercialised end of children's media. My final two examples suggest that public service media companies may also be engaging in similar attempts to reach the global market—in both cases, of pre-school children.

The US series *Sesame Street* is now more than 35 years old. Produced by a not-for-profit company, and screened in the US on public television, its success and survival have crucially depended on global marketing. Just as with Pokémon, there are global Sesame Street franchises, and an enormous range of spin-off merchandise—although because of its educational cachet, these tend to be the kinds of products middle-class parents are less likely to object to.

Interestingly, *Sesame Street* was also the subject of a forceful ideological critique from Armand Mattelart, the co-author of *Reading Donald Duck* (Mattelart and Waksman, 1978). He saw *Sesame Street*, like Disney, as the bearer of a set of US ideological values, and as part of a broader cultural imperialism. Just as Disney's appeal to childhood allowed it to profess an essential innocence, he argued, so the educational intentions of *Sesame Street* served as a kind of alibi. According to Mattelart, the programme was actually putting across a specific set of social values and capitalist ideologies that children were assumed to simply swallow whole. Perhaps surprisingly, this argument was also made in the UK as well, albeit in slightly different terms. There was a great controversy in the late 1960s when the BBC refused to buy *Sesame Street*, not so much on ideological grounds as on the basis of its style of pedagogy. The use of what were seen as 'advertising techniques' for drilling children in letter and number recognition was seen as somehow at odds with the more child-centred, play-oriented British tradition of pre-school education (see Buckingham *et al.*, 1999).

In fact, however, *Sesame Street* has frequently worked with local producers around the world to insert local content. Alongside Big Bird and Elmo and Kermit, the programmes typically include locally-produced documentary inserts that feature children in the country where the programme is screened. There has also been some research looking at how, for example, *Sesame Street* in Israel incorporated material about Israeli and Palestinian children, and the consequences of this in terms of their attitudes towards each other (Fisch, 2001).

Ironically, in recent years the BBC has been engaged in a very similar enterprise. Its ground-breaking pre-school series of the late 1990s, *Teletubbies*, was clearly produced with a global audience in mind; and indeed, it has become increasingly imperative for the BBC to earn revenue from overseas sales, even though it is primarily funded by compulsory taxation (the license fee). An investment on the scale of *Teletubbies* would not have been possible without assured overseas revenue, not just from programme sales but also from ancillary merchandising—and, as with the other texts I have discussed, the scale and diversity of the merchandising is quite phenomenal (Buckingham, 2002).

Like *Sesame Street*, *Teletubbies* also offers the facility for local broadcasters to insert local content. For example, despite resistance on the part of the Norwegian public broadcasters to buying it (on grounds that were not dissimilar to the BBC's concerns about *Sesame Street*), the programme is now screened in Norway, and includes not just Norwegian voice-overs and dubbing but also documentary material featuring Norwegian children. Again, this could be seen as an example of glocalisation, a productive meeting of the global and the local. Nevertheless, there are questions to be raised about what is and is not culturally specific here. It may be that cultural specificity is not simply a matter of content, but also of form—or in educational terms, it is not just about the curriculum of the programme, but also about its pedagogy. *Teletubbies* has a very different style of pedagogy compared with *Sesame Street*; and while I would hesitate to label one essentially British and the other essentially 'American', there clearly are cultural differences here that would be worth investigating more closely.

## Conclusion: Harry Potter Goes Global

The forms that this 'glocalisation' takes—and its consequences for children's sense of cultural identity—are therefore diverse and variable. The notion of cultural imperialism does to some extent describe what is taking place here, at least at an economic level; but it fails to account for the diversity and complexity of how children use and interpret cultural texts. On the other hand, there are reasons to be cautious about the postmodern emphasis on hybridity and fragmentation: access to global markets is not equally open to all, and consumers are clearly not free to choose their cultural identities from an infinite range of global possibilities.

Yet the global marketing of children's culture constantly throws up new paradoxes. The current worldwide success of Harry Potter is a striking case in point. Harry Potter seems to me at least to be highly culturally specific. It is distinctly British, indeed essentially English: it draws on a middle-class British tradition of school stories, although it could also be seen to reflect some rather more contemporary concerns and anxieties in British culture (Blake, 2002). Nevertheless, it clearly has global appeal: like the other phenomena I have discussed, it is effectively a global brand, which is used to sell a whole range of media and merchandise—books, films, computer games, posters, toys, clothing, gifts and paraphernalia of all kinds—to children around the world.

The reasons for this remain to be established. How, for example, do we explain the success of Harry Potter in China? What it is that Chinese children, whose lives are very different in many ways from those of children in Britain, seem to recognise in Harry Potter? Does this point to the existence of some kind of universal, global childhood—or is it that these texts are interpreted in such different ways in different cultures that they effectively become very different things? Does Harry Potter, for all the apparently old-fashioned nature of the story, somehow represent a *modernist* conception of childhood, which transcends cultural differences?

And so I conclude with a series of questions. Is Kenichi Ohmae correct: is global marketing really creating a common culture of childhood? Is it helping children to communicate across cultural differences—or is it simply eradicating those differences? And even if it is, is that something we should necessarily regret?

### Note

1.  To some extent, the reverse is also true, but there is also a strong aspirational element at work here: children frequently aspire to consume things that seem on the face of it to be targeted at a somewhat older audience. This is particularly an issue for children in the immediate pre-teen age group, for whom the category of the 'teenager' is seen to embody a degree of freedom from adult constraints (de Block, 2000).

### References

Allison, A. (2000) A challenge to Hollywood: Japanese character goods hit the US, *Japanese Studies*, 20(1), 67–88.

Barker, M. (1990) *Comics: Ideology, Power and the Critics*, Manchester: Manchester University Press.

Bennett, A. (2000) *Popular Music and Youth Culture*, London: Palgrave Macmillan.

Blake (2002) *The Irresistible Rise of Harry Potter: Kid-Lit in a Globlaised World*, London: Verso.

Buckingham, D. (2000) *After the Death of Childhood: Growing Up in the Age of Electronic Media*, Cambridge: Polity.

Buckingham, D. (2001) Disney dialectics: debating the politics of children's media culture, in J. Wasko, M. Phillips and E. Meehan (eds) *Dazzled by Disney*, London: Leicester University Press, 269–96.

Buckingham, D. (2002) Child-centred television? *Teletubbies* and the educational imperative, in D. Buckingham (ed.) *Small Screens: Television for Children*, London: Leicester University Press.

Buckingham, D. (2004) *Young People and Media*, Briefing paper for the United Nations Workshop on Global Media-Driven Youth Culture, New York.

Buckingham, D. and Sefton-Green, J. (2004) Gotta catch 'em all: structure, agency and pedagogy in children's media culture, in J. Tobin (ed.) *Pikachu's Global Adventure: The Rise and Fall of Pokemon*, Durham, NC: Duke University Press, 12–33.

Buckingham, D., Davies, H., Jones, K. and Kelley, P. (1999) *Children's Television In Britain: History, Discourse and Policy*, London: British Film Institute.

Burton-Carvajal, J. (1994) 'Surprise package': looking southward with Disney, in E. Smoodin (ed.) *Disney Discourse: Producing the Magic Kingdom*, New York: Routledge, 131–47.

Cartright, L. and Goldfarb, B. (1994) Cultural contagion: on Disney's health education films for Latin America, in E. Smoodin (ed.) *Disney Discourse: Producing the Magic Kingdom*, New York: Routledge, 169–80.

Cohen, P. (1997) *Rethinking the Youth Question*, London: Macmillan.

Cowen, T. (2002) *Creative Destruction: How Globalisation is Changing the World's Cultures*, Princeton, NJ: Princeton University Press.

de Block, L. (1998) From childhood pleasures to adult identities, *English and Media Magazine*, 38, 24–9.

Dorfman, A. and Mattelart, A. (1975) *How to Read Donald Duck: Imperialist Ideology in the Disney Comics*, New York: International General.

Drotner, K. (2001) 'Donald Seems So Danish': Disney and the formation of cultural identity, in J. Wasko, M. Phillips and E. Meehan (eds) *Dazzled by Disney*, London: Leicester University Press, 102–20.

Feld, S. and Keil, C. (1992) *Music Grooves*, Chicago, IL: University of Chicago.

Featherstone, M. (1995) *Undoing Culture*, London: Sage.

Fisch, S. (2001) *G is for Growing: Thirty Years of Research on Children and;Sesame Street*, Mahwah, NJ: Erlbaum.

Forgacs, D. (1992) Disney animation and the business of childhood, *Screen*, 33(4), 361–74.

Gillespie, M. (1995) *Television, Ethnicity and Cultural Change*, London: Routledge.

Giroux, H. (2001) *The Mouse that Roared: Disney and the End of Innocence*, New York: Rowman and Littlefield.

Griffiths, M. (2002) Pink worlds and blue worlds: a portrait of infinite polarity, in D. Buckingham (ed.) *Small Screens: Television for Children*, London: Leicester University Press.

Hannerz, U. (1996) *Transnational Connections: People, Culture, Places*, London: Routledge.

Hendershot, H. (2004) *Nickelodeon Nation: The History, Politics and Economics of America's Only TV Channel for Kids*, New York: New York University Press.

Iwabushi, K. (2004) How 'Japanese' is Pokémon?, in J. Tobin (ed.) *Pikachu's Global Adventure: The Rise and Fall of Pokémon*, Durham, NC: Duke University Press, 53–79.

James, A., Jenks, C. and Prout, A. (1998) *Theorizing Childhood*, Cambridge, Polity.

Katsuno, H. and Maret, J. (2004) Localizing the Pokémon TV series for the US market, in J. Tobin (ed.) *Pikachu's Global Adventure: The Rise and Fall of Pokémon*, Durham, NC: Duke University Press, 80–107.

Kline, S. (1993) *Out of the Garden: Toys and Children's Culture in the Age of TV Marketing*, London: Verso.

Leshkowich, A. and Jones, C. (2003) What happens when Asian chic becomes chic in Asia?, *Fashion Theory*, 7(3–4), 281–99.

Liebes, T. and Katz, E. (1990) *The Export of Meaning*, Oxford: Oxford University Press.

Livingstone, S. (1998) Mediated childhoods; a comparative approach to young people's changing media environment in Europe, *European Journal of Communication*, 13(4), 435–56.

Livingstone, S. (2002) *Young People and New Media*, London: Sage.

Lustyik, K. (2003) *The Transformation of Children's Television from Communism to Global Capitalism in Hungary*, Boulder, CO: University of Colorado.

McGray, D. (2002) Japan's gross national cool, *Foreign Policy*, May–June.

McQuillan, M. and Byrne, E. (1999) *Deconstructing Disney*, London: Pluto.

Mattelart, A. and Waksman, D. (1978) *Plaza Sezamo* and an alibi for the author's real intentions, *Screen Education*, 27, 56–62.

McChesney, R. (2002) Media globalisation: consequences for the rights of children, in C. von Feilitzen and U. Carlsson (eds) *Children, Young People and Media Globalisation*, Goteborg, Sweden: UNESCO International Clearinghouse on Children, Youth and Media, 33–42.

Ohmae, K. (1995) *The End of the Nation State*, New York, Harper Collins.

Piedra, J. (1994) Pato Donald's gender ducking, in E. Smoodin (ed.) *Disney Discourse: Producing the Magic Kingdom*, New York: Routledge, 148–68.

Punch, S. (2003) Childhoods in the majority world: miniature adults or tribal children?, *Sociology*, 37(2), 277–95.

Robertson, R. (1994) Globalisation or glocalisation?, *Journal of International Communication*, 1(1), 33–52.

Roth, M. (1996) A short history of Disney fascism, *Jump Cut*, 40, 15–20.

Schneider, C. (1992) *Children's Television*, New York: Contemporary Books.

Silj, A. (1988) *East of Dallas: The European Challenge to American Television*, London: British Film Institute.

Singer, D. and Singer, J. (2002) *Handbook of Children and the Media*, New York: Sage.

Smoodin, E. (ed.) (1994) *Disney Discourse: Producing the Magic Kingdom*, New York: Routledge.

Tobin, J. (ed.) (2004) *Pikachu's Global Adventure: The Rise and Fall of Pokémon*, Durham, NC: Duke University Press.

Tomlinson, J. (1991) *Cultural Imperialism*, Baltimore, MD: Johns Hopkins University Press.

Tunstall, J. (1977) *The Media Are American*, London: Constable.

van Evra, J. (2004) *Television and Child Development*, Mahwah, NJ: Erlbaum.

von Feilitzen, C. and Carlsson, U. (eds) (2002) *Children, Young People and Media Globalisation*, Goteborg, Sweden: UNESCO International Clearinghouse on Children, Youth and Media.

Wagnleitner, R. (1994) *Coca-colonisation and the Cold War: The Cultural Mission of the US in Austria After the Second World War*, Chapel Hill, NC: Duke University Press.

Wasko, J., Phillips, M. and Meehan, E. (eds) (2001) *Dazzled by Disney*, London: Leicester University Press.

Westcott, T. (2002) Globalisation of children's TV and the strategies of the 'big three', in C. von Feilitzen and U. Carlsson (eds) *Children, Young People and Media Globalisation* Goteborg, Sweden: UNESCO International Clearinghouse on Children, Youth and Media, 69–76.

Yano, C. (2004) Kitty litter: Japanese cute at home and abroad, in J. Goldstein, D. Buckingham and G. Brougere (eds) *Toys, Games and Media*, Mahwah, NJ: Erlbaum.

# Hungry Children and Networks of Aid in Argentina: Thinking About Geographies of Responsibility and Care

FERNANDO J. BOSCO
*Department of Geography, San Diego State University, USA*

## Introduction: Children's Geographies and Geographies of Responsibility and Care

One of the most significant effects of globalization in the context of neo-liberal regimes has been a restructuring and rescaling of welfare provision (Herod and Wright, 2002; Staeheli and Brown, 2003). The rise of local, national, and global non-government organizations (NGOs) has been taken as an indication that responsibility for protection and for the rights of people (including children, of course) is shifting from government institutions to governance networks that operate at different scales (Salamon and Anheier, 1999). As Dyck (2005, p. 239) has argued, the combining effects of globalization and neoliberalism have resulted in the commodification and rescaling of welfare provision and given rise to complex chains of care. Such changes have puzzling social and spatial consequences. For example, it is still unclear how responsibilities regarding welfare provision are defined through these new trans-local networks of people, organizations, and institutions. There are, moreover, unanswered questions regarding the scope and effectiveness of such com- modified, rescaled, and networked strategies of welfare. How is caring for and about vul- nerable others (e.g., the elderly, children) negotiated and mediated across actors and scales? And more importantly, despite good intentions, how effective are these strategies in helping and caring for those in need?

These questions are relevant to work in children's geographies because they speak to spatial dimensions of responsibility and care in relation both to children's livelihoods and the social construction of children. Questions regarding the welfare of children can be positioned in relation to geographic research that deals with care and caregiving activi- ties—from the provision of healthcare to the family care provided by different types of caregivers even to the emotional work involved in a caregiving relationship (Mitchell *et al.*, 2004; Dyck, 2005). In addition, questions about caring for children can be positioned relative to research on ethics and geography—or what some have called 'moral geographies' (Smith, 1997).[1] In this area, geographers recently have examined issues of responsibility (Corbridge, 1993; Smith, 1998; Massey, 2004) and hospitality towards others (Barnett, 2005), aiming to develop investigations of spatial patterns and relations that invite a moral reading (see Smith, 1997, p. 585). As geographers begin to engage research on children and young people with wider conceptual and intellectual concerns (see for example Aitken, 2001; Katz, 2004), there is ample opportunity to develop research that examines the relations between care, ethics and children. Recently, there have been calls for children geographers to engage more aggressively with new ideas from human geography that might '... prove useful, enabling, thought provoking, constructive, telling or interesting in apprehending ... children's lives' (Horton and Kraftl, 2006, p. 70). An engagement of children's geographies with some of the ideas that geographers are articulating in relation to caring and responsibility seems in order.

In this paper, I connect current discussions of geographies of responsibility and care with contemporary geographic work that, as Aitken (2004, p. 579) explains, places children '... at the center of our understanding of consumption, production and reproduction, and at the heart of the inequities generated by globalization'. I want to make these connections because I am interested in relating more abstract discussions about the spatial scope of responsibility and care to the very concrete issue of the delivery of humanitarian aid to children. In this context, I am interested in examining how caring for and helping children is increasingly taking place through various forms of networks that link people and different organizations and institutions of civil society. I make this connection by situating my discussion empirically in the context of the economic, social, and political crisis that has affected Argentina since 2001 and that has had a devastating toll on the lives of children and young people in the country. Specifically, I discuss the impacts of this crisis on the welfare of children and young people and highlight some of the responses that emerged from different grassroots efforts as people tried to care for hungry and sick children in Argentina. My main objective in this paper is to critically interrogate new modes of care-giving with specific reference to poor and needy Argentine children. I offer a narrative that problematizes the effectiveness of networks of solidarity and aid and the search for responsibility in such networks. My intention is not to demonize the real and praiseworthy *intentions* of those people who are trying to care for children and young people who have been abandoned by the traditional functions and institutions of the welfare state. Rather, I want to offer some critical insights into the limitations and/or advantages of some approaches to the delivery of aid and care for children that have been used in Argentina and elsewhere.

## Helping/Caring for Children: Positioning the Argument

I am specifically interested in discussing and assessing the effectiveness of aid and care for children that is mobilized through internet-based networks. Since the 1990s, enhancing and improving access to the internet around the world (e.g., through the use of electronic mail and the world wide web) and taking advantage of advanced information technology has been touted as one effective approach to overcome logistical and organizational challenges of development in the so-called Global South—witness calls to use information technologies universally for capacity building or to bridge the digital divide (Taylor, 2004).[2] Many have argued that the expansion and increased use of the internet and world wide web around the world would result in new virtual communities that would generate new forms of political engagement and participation (Rheingold, 1993; Tapscott, 1998). Even the more critical social movement literature has been peppered with case studies of the successful mobilization of activists via the internet, the emergence of cyber-cultural politics, and the practice of political activism at a distance (Ribeiro, 1998). Some of the overly utopian and optimistic views regarding the transformative power of the internet and cyber-cultural politics have already been challenged (see for example Robins and Webster, 1999). However, we still know very little about the effectiveness and effects of internet-based approaches to the delivery of aid and care for others. How effective are they over space and time? Do they encourage responsible participation? How do distant others understand the very specific needs of local populations? Are those responsible for putting together these networks really understanding those needs in the first place?

These are broad questions that require empirical investigation and further elaboration. In this paper, I approach such questions through a critical assessment of the effectiveness of one specific internet-based NGO that works to aid and care for needy children in Argentina. This organization is called *Por Los Chicos* (For the Kids). The main goal of this NGO is to combat hunger among needy Argentine children. The NGO distributes food to other NGOs and institutions that work with poor children in Argentina. *Por Los Chicos* uses the internet to collect donations from Argentina and beyond. The main characteristic of this NGO (which I describe in more detail later in this paper) is its embrace of new communications technologies to build a global network of aid for Argentine children. My critical assessment of the approach followed by this organization goes beyond an initial understanding of effectiveness that is based on measures of aid delivered (such as quantities of food distributed, amount of money donated, and so on). While I analyze the overall results of the network of aid in question, my goal is to interrogate critically whether the development of the network itself has led to the creation of a more effective geography of care among those who participate. Specifically, I discuss how enrollment of donors in the network of *Por Los Chicos* is related to particular representations of 'hungry' and 'needy' children. I am interested in discussing the effects that the mobilization of particular representations of children has on the overall effects of networks of care-giving. As a geographer, I am interested also in examining what takes place when representations of children became unhooked from their local context and began circulating in trans-local networks of aid and solidarity.

I use my assessment of this internet-based organization as a way to speak to the challenges we face when we attempt to care for children at a distance and to the effectiveness (or limitations) of the shift to more decentered and networked forms of governance that characterize the contemporary landscape of care and welfare provision. I also contrast my critical assessment of the internet-based network of aid for children with the work of another network of aid that operates trans-locally from Argentina: the *Red Solidaria* (Solidarity Network). I aim to position the work of the *Red Solidaria* in relation to some ideas about care that are emerging in critical human geography, inspired by feminist work on critical ethics of care and by relational notions of responsibility. I argue that some of these ideas might provide us with a platform from which to begin a more critical analysis of the effectiveness of networked forms of aid and welfare, in particular, if what is at stake is the valuing of humans and the creation of attachments to others.

Rather than attempting to reach a grand theoretical synthesis of all relational approaches to caring for and on behalf of others, I pursue a more pragmatic and normative approach. In this paper, my modest goal is to offer some ideas that can be applied not to the question of why we should care or how far we should care about others (questions some geographers have already tackled, see Smith, 1998 and Barnett, 2005) but rather to the question of *how best to care and act responsibly about others*—whether those others are close or not so close to us (on this issue, see also Wenar, 2003). I argue that this question is important because it speaks to whether the strategies of care that get implemented are effective or ultimately fail.

## Argentina's Twenty-first Century Economic and Social Crisis in Context: Children in Poverty

In December 2001, Argentina plunged into the biggest economic crisis of the country's history. Many analyses have seen the crisis as the result of the failure of the neo-liberal prescriptions that the state had been pursuing (with the support and encouragement of multinational lending agencies) for the previous two decades. The 1990s were years of extreme changes in the organization of the Argentine economy and society. These changes were based on a complete restructuring of the Argentine economy and state according to a neo-liberal approach. These changes demanded the dismantling of the welfare state, the privatization of state owned enterprises and the complete liberalization of the economy. For example, following a neo-liberal approach to development common to most Latin American countries in the 1990s, the majority of Argentina's public companies (utilities, transportation, energy generation, even the postal service) had been privatized. Argentina embraced principles of free trade and abandoned any kind of protection for local industry, small businesses and entrepreneurs (Bosco, 1998). The market was flooded with imported goods, and people in Buenos Aires had access to the latest global products. Shopping at *Galerias Pacifico* in Buenos Aires (a new mall in a nineteenth century building that once housed the administration and operation center of the now defunct national railroads) was like shopping in any large mall in the United States. Both the conditions and structures of production and consumption were radically altered. At the same time, and following the prescription of structural adjustment plans that were part of neo-liberal ideology and the 'Washington Consensus', the Argentine state had done everything possible to become 'lean' and to drastically reduce all kinds of public expenditures. For example, besides the privatization of all state-owned enterprises, people were encouraged to switch their retirement savings from the public funded retirement system into a new, private system. Reductions in funding for education, public health and other social services became the norm, leading to a drastic transformation of the welfare state in Argentina.

The consequences of this transformation were widespread. While workers were being convinced to leave their retirement savings to the workings of the market and the promise of higher returns on their investment, something else was happening to Argentine society. Income disparity reached historic highs, and the unemployment rate was above 20%. Local industries were unable to compete because of higher labor costs and lack of any kind of national or regional industrial policy. Since the state abandoned almost all of its roles in the provision of social welfare, there was rapid growth in the private sector provision of services such as education, housing and health (Bosco, 1998). This meant that those in the lower strata of society became more vulnerable to economic fluctuations and the restructuring of production and labor markets. Income gaps widened and the apparent affluence seen in downtown Buenos Aires was in stark contrast to the rampant poverty affecting the outskirts of the city and many other urban and rural areas. Internal consumption became stagnant and industrial production was virtually non-existent. The money that the state had acquired from privatization schemes ran out as the government attempted to service ballooning interest payments on international loans taken throughout the 1970s and 1980s. The neo-liberal recipe failed the people, except those few who were connected to the circuits of global capital and were benefiting from the new approach. These people were in the minority of the population.

The arrival of the twenty-first century saw Argentina increasingly unable to cope at once with internal demands for job creation and general welfare and the external demands for further restructuring in order to fulfill debt and loan obligations. By 2001, more than 50% of people in the largest urban areas of the country were living in poverty (in some urban centers that figure reached 70%) (Figure 1), and over 30% of those people were living in what is considered by the Argentine Institute of Statistics and Census to be 'extreme poverty' (Figure 2). At the end of 2001 the government decided to freeze all bank accounts indefinitely and devalue the currency, as the economy was faltering and Argentina was unable to meet financial obligations. There were limits on the amount of money people could withdraw from their bank accounts, further squeezing the middle classes (North and Huber, 2004). A popular uprising involving mass demonstrations and brutal police repression of civilians led to the resignation of the president and to months of political instability and social unrest. Food riots and popular demonstrations occurred weekly in

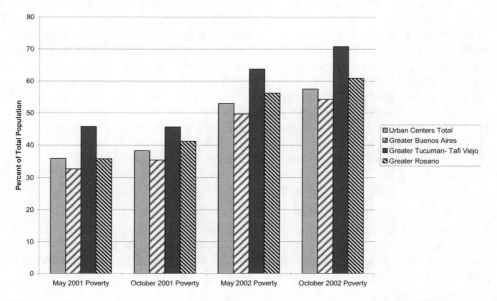

**Figure 1.** Argentina—Percentage of Population Living in Poverty, 2001–2002 (by major urban areas). Data source: *Instituto Nacional de Estadísticas y Censos*, Argentina.

*Global Childhoods*

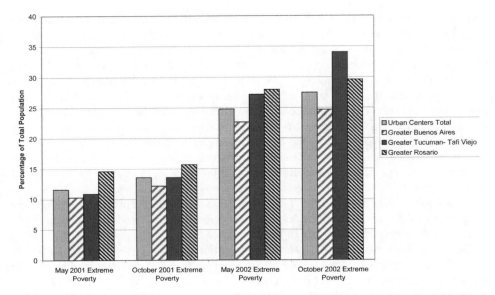

**Figure 2.** Argentina—Percentage of Population living in conditions of extreme poverty, 2001–2002 (by major urban areas). Data source: *Instituto Nacional de Estadísticas y Censos*, Argentina.

major cities across the country. Middle class people went out on the streets, banging pots and pans, demanding that the government return their money. Poor people struggled to find food and feed their children. In the first two months of 2002, the Argentine GDP sank by 16.3% and manufacturing output was reduced by 20% (North and Huber, 2004). The country was bankrupt and Argentina declared itself in default. It became the largest sovereign default in history. The neo-liberal experiment had gone terribly wrong.

In the context of this crisis, Argentina's children suffered the most. Unfortunately, their suffering was not new. Income disparity and poverty had been increasing in the country for several years before the crisis. By the time the economic crisis erupted, children in poor and marginal areas had long been depending on milk, bread and soup given to them at public schools for their daily meals. Yet, many in Argentina had consistently ignored this reality and instead focused their attention on touting the benefits that Argentina was accruing from its embrace of neo-liberalism. However, right after the collapse of the Argentine economy, the national and world media began reporting on the social costs of the crisis on an almost daily basis. For example, in an article entitled 'Argentina: Life after bankruptcy' published in September 2002, the French newspaper *Le Monde Diplomatique* described the situation of Argentine children in very accurate, if gruesome, terms:

> Young people have been showing visible malnutrition for two years and the situation has worsened in recent months . . . hungry children are fainting; absenteeism at school is down since primary school children do not want to skip the food offered at school, which is often their only meal of the day. Sometimes mothers appear at schools with empty plates, demanding food for sick children at home. Earlier this year, this was happening only in the most impoverished province of Tucumán, now it happens nationwide, including in Buenos Aires province, where for the first

time 100 schools kept their cafeterias open over the winter . . . Children in the town of Quilmes, only 30 km from the capital, are reduced to eating fried toads or even rats. (Auge, 2002)

Children became the main actors of a national tragedy that received worldwide attention. Everyone now knew that for the first time in modern Argentine history, children were dying of malnutrition, starvation and poor health in one of the most food- and resource-rich countries in the Western Hemisphere. Data from 2003 indicates that about 62% of children were living below the poverty line in all major urban centers in Argentina (Figure 3); 30% of those children were living in conditions of extreme poverty (Figure 4). In the case of the city of Tucumán, almost 80% of all children were living in conditions of poverty; and 30% of those children were living in conditions of extreme poverty (Figure 4). Rosario, Argentina's second largest city, also had 60% of children living below the poverty line, and more than half of those children were living in conditions of extreme poverty (Figure 4). The majority of children's deaths that occurred as a result of malnutrition during the economic crisis took place in these two cities.

The percentages of poor children decreased some by 2004, when Argentina's economy began to recover and the distribution of welfare improved as a result of bigger government revenues and additional funds dedicated to welfare and poverty alleviation. Today, the statistics are still startling: 60% of children in all urban centers are still living under conditions of poverty (Figure 3) and over 25% of those are living in conditions of extreme poverty (Figure 4). Cities in the interior of the country (such as Tucumán and Rosario, where most of the cases of malnutrition were documented) still presented some of the largest percentages of children living in poverty and extreme poverty in 2004 (Figures 3 and 4).

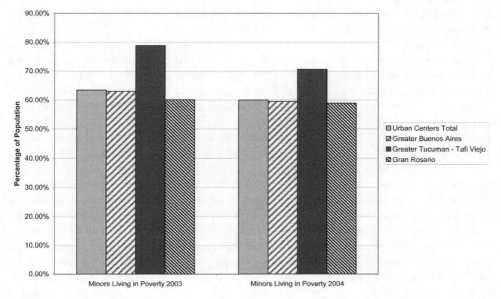

**Figure 3.** Percentage of Children in Argentina living below the poverty line, years 2003 and 2004 (all major urban centers and selected urban centers). Data source: *Instituto Nacional de Estadísticas y Censos*, Argentina.

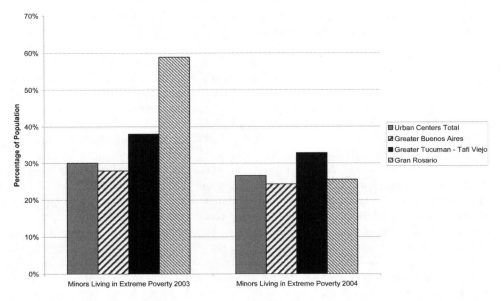

**Figure 4.** Percentage of Children in Argentina living in conditions of extreme poverty, years 2003 and 2004 (all major urban centers and selected urban centers). Data source: *Instituto Nacional de Estadísticas y Censos*, Argentina.

## Hungry Children, the Internet, and Networks of Aid

As a result of the inability of the state to cope with a human crisis that had unprecedented effects in the country, new networks of aid and advocacy emerged that grouped together non-profit and non-governmental organizations (NGOs). The new NGOs continued the development of an 'associational revolution' in global civil society (Salamon and Anheier, 1999) that has characterized the development field in Latin America in recent decades (Keck and Sikkink, 1998). In other words, the NGOs and networks of advocacy and aid were civil society's response to the economic crises and to the consequences of structural adjustment programs. Many of the NGOs were committed to raise and distribute supplies of food, clothes, medicine and other needed items to impoverished populations, in particular children. In attempting to provide basic human welfare functions that the state abandoned or was unable to provide, the networks began coordinating the aid contributions of people and other organizations that operated both in Argentina and abroad.

Individual donations by people located in different countries became important because, as a result of the economic crisis, many Argentines also began migrating out of the country and relocating to Europe and the United States (the United States and Spain were among the preferred destinations). These Argentines abroad soon formed expatriate communities in their new host societies. The new NGOs and networks of aid tapped into this Argentine diaspora in the Global North for donations and assistance. Many in the expatriate community also felt compelled and obliged to do something about the appalling conditions in Argentina and found the solidarity networks a good outlet for their beneficence. Many of these NGOs functioned over the internet as networks linking people committed to help, and were created by Argentines themselves, whether in Argentina or abroad. Examples of these networks include *Help Argentina* (funded by a partnership between an American and an Argentine and working both in Argentina and the United States), *Red Solidaria* (based in Buenos Aires and operating locally across Argentina), and

*Por Los Chicos* (based in Buenos Aires but operating transnationally through the world-wide-web), among others.

Of all these new NGOs, *Por Los Chicos* received the most attention in the media and was touted as an example of the success of grassroots efforts in bringing about help for needy children and young people in the context of the crisis. *Por Los Chicos* describes itself as an NGO that works to combat hunger among needy children in Argentina.[3] The NGO's main activities are supported by their use of the internet to link donors and sponsors in a large network of aid. Their internet-based network approach is based on the model of the 'Hunger Site' and other similar websites based in the United States. These websites rely on the popularity of the notion of 'corporate social responsibility' among for-profit corporations and enterprises. *Por los Chicos* recruits corporate actors (and even some other non-profit organizations) to acts as sponsors for their website. These corporate sponsors support the activities and projects of the NGO by advertising on the website. In the case of *Por Los Chicos*, each time a visitor to the website clicks on a banner that displays the words 'Donate Free Food', the new donor is faced with an advertisement for a particular sponsoring corporation or organization. Thus, the amounts of donations given to the NGO is dependent of the number of times people visit the site and click on banners that take them to advertising that is sponsored by 'socially responsible' corporations. Each click of the mouse or keyboard on a 'Donate Free Food' banner and its corresponding advertising results in a certain amount of money donated to the NGO, which is then used to buy food that is distributed to feed those children in need across Argentina.

The idea that drives *Por Los Chicos* is appealing, and the data available through the organization website indicates that at the beginning of the Argentine crisis, the network received a lot of support, both by sponsors and donors, and that this translated into help for children in Argentina. For example, in 2002, there were a total of 33 sponsors for the website, and over 3.5 million donations, which translated into the equivalent of 1 million meals for children. However, examination of the data over time seems to indicate that the effectiveness of such internet-based approach to the delivery of aid was short-lived. Even though a year later, in 2003, the number of donors (visitors to the site) increased to over 6 million, the total number of meals donated decreased to just over 800,000 as the number of sponsors decreased to 28. In 2004, the number of sponsors decreased again (to 16), and so did the number of donations, resulting in about 460,000 donated meals, almost 50% less than a year earlier. In 2005, the total number of sponsors decreased to just 13 and there were months during that year when only one or two donors were willing to sponsor the NGO for food donations. For example, in May of 2005 there was only one sponsor listed in the *Por Los Chicos* website, and interestingly enough, the sponsor was another NGO formed by expatriate Argentines living in Seattle, USA. The data available for 2006 indicates that the NGO has only been able to attract a total of seven sponsors, and several of them have acted as donors for just one month. After more than fours years of activity, the number of visitors to the site has also dwindled and the number of meals donated decreased substantially to just over 250,000.[4] Yet, about 60% of children in Argentina were still living in conditions of poverty and 25% of all children were living under conditions of extreme poverty in 2005 (Figure 4). The crisis facing children in Argentina has not subsided, but the effectiveness of the web-based NGO in attracting donors and visitors has decreased substantially.

Closer inspection of data on 'donors' and 'donations' seems to confirm that the model that the NGO adopted is not sustainable over the long term. It is not only that the private sponsors who are responsible for donating funds to buy food gradually abandoned the network (perhaps as the media stopped reporting on the crisis because the economy in

*Global Childhoods*

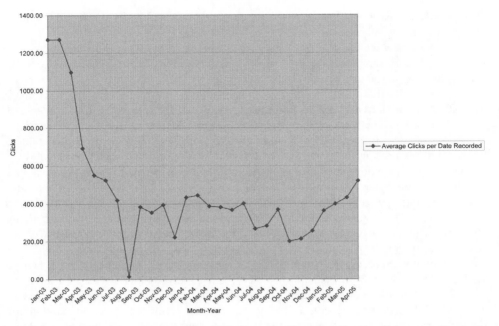

**Figure 5.** Average monthly number of clicks in the 'Donate Free Food' banner of a single sponsor in the *Por Los Chicos* website, January 2003 to May 2005.

Argentina began to recover), but also that visitors to the site (those who generate the donations in the first place by clicking on the *Donate Free Food* banners) also diminished over time. Figure 5 shows the average monthly number of clicks on the *Donate Free Food* banner that has been connected to the sponsorship of another organization called *Argentines in Seattle* (a group of expatriates living in the United States) that has been an active supporter/donor of the network. As Figure 5 indicates, the number of clicks on their banner has showed a consistent decline, going from an average of about 1300 'clicks' at the beginning of 2003 to a low of 200 'clicks' by the end of 2004. Similarly, Figure 6 shows the monthly average number of contributions given to *Por Los Chicos* by this group of Argentines living in Seattle over the same period of time. As Figure 6 shows, the average amount of donations has also declined since 2003, mostly because contributions are tied to the number of visitors to the main *Por Los Chicos* website and the number of clicks on the advertising banners.

These data are limited because they represent a snapshot of clicks and donations recorded by just one donor. However, it confirms the declining trend that was described above and that was based on the total number of visitors and donations to the website. Evidence of the declining numbers of sponsors and donors and amounts of donations indicate the diminishing impact of this internet-based approach in helping children over time. Overall, the efforts of the NGO, while praiseworthy, have not been completely successful because it has been unable to sustain interest in, and support for, its mission.

The NGO faces severe limitations that, I argue, relate to the inability of this particular approach to engage both donors and sponsors in a meaningful manner with the plight of child hunger and malnutrition in Argentina. These problems are related to the broader issue of the limitations of modern communication technologies in creating a meaningful sense of involvement among distant others. Geographer David Smith, for example, has argued that while modern technologies such as the internet can motivate practical care,

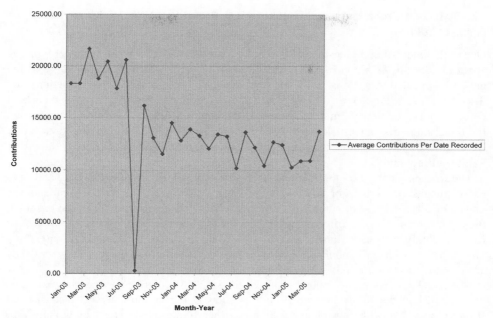

**Figure 6.** Average monthly amount of contributions by a single sponsor of the *Por Los Chicos* website, January 2003 to May 2005.

they might also induce numbness, indifference, or even enable horrors to be made picturesque (Smith, 1998, p. 22). As Smith puts it:

> People engage possibly far distant others in ways usually associated with close proximity, but without face-to-face interaction or bodily contact . . . however, these virtual communities are a luxury of well-to-do members of affluent societies, whose scope for caring relationships via the internet is confined to others with similar resources. They may therefore do nothing more than reinforce privilege. Furthermore, there are some relationships of mutuality and care which actually require knowledge of the other as physically embodied, capable of touch, which cannot be regulated by the choice to switch on the computer. (Smith, 1998, p. 22)

Smith's argument about spatial relationships of care through modern media such as the internet relates to the inability of participants in the *Por Los Chicos* network to meaningfully engage with the problem of hungry children in Argentina. Additionally, I argue that in this specific case, the problem is further compounded by the way in which the NGO and the internet-based network model was imagined and assembled when it was created. For example, the motivations behind the creation of the web-based NGO were questionable from the beginning, since its founder and director (who is also a banker associated with a large transnational bank based in the United States) publicly explained to the BBC News that the website was helping fight the growing anti-capitalist feeling that spread throughout Argentina after the country's economic crisis by demonstrating that private companies wanted to help (Hale, 2002). Beyond the suspicious association of the NGO with some of the stakeholders and the ideology that had contributed to the Argentine economic crisis in the first place, I argue that the problem facing *Por Los Chicos* is also related to the ideas about responsibility and the obligations to needy children that were used to justify the existence and functioning of the network in the first place.

### The Universal Child, Universal Notions of Care, and Symbolic Violence against Children

One of the interesting dimensions of the formation of this NGO was the process of *talking* about children to *help children,* and the different ways in which those in charge of the network appealed to a global audience that already buys into specific constructions and discourses of childhood in the developing world. Indeed, *Por Los Chicos* mobilizes certain images of needy children to appeal to donors at multiple scales. Even though the images and discourses used are based upon real material conditions (children *really* were hungry and dying), the constructions of children mobilized by the network are totalizing and problematic.

The images featured on the main web site are completely decontextualized, and only show images of hungry or sad children, apparently staring at the camera and the viewer. Each time a visitor reloads or re-visits the website from a computer terminal, the main screen shows a different image of a lonely, hungry child, either with a plate of food or simply passively staring, seemingly waiting to be fed and helped (Figure 7). A visitor to the website cannot tell anything about these children (who and where they are, what their lives are like, what has happened to their families, and so on).

The NGO's iconography follows what Ruddick (2003) describes as the dominant representational model of the universal child of developing nations, always disconnected from context and with few clues as to his or her background. The iconography also builds on an aesthetic of repetition which produces a particular kind of generalization (see Aitken, this volume). Thus, in the case of *Por Los Chicos*, the construction of aid and solidarity networks for needy children becomes dependent on the use of decontextualized and repetitive imagery of hungry children as seductive symbolic and rhetorical devices.

**Figure 7.** The main *Por Los Chicos* opening website as seen from a computer screen. The image of the hungry child dominates the opening screen.

Previous research on the importance of iconography has indicated that images of suffering are often appropriated to appeal emotionally to different audiences. There is extensive documentation that demonstrates that the print and audio visual media are often complicit in this process (see Moeller, 1998). In the process of reaching out to others to help children in need, *Por Los Chicos* became part of a development and aid industry that has long relied on the use of decontextualized images as the symbolic repertoire to appeal emotionally to donors at different levels. But as Moeller (1998) argues, the regular use of images of suffering, pain, hunger or need often leads to 'compassion fatigue'. Moeller describes compassion fatigue as a multifaceted process in which images of suffering are appropriated and become commercialized by the media. Suffering becomes another commodity. Faced with a visual avalanche of pain and suffering, audiences (readers, viewers, and in the case of *Por Los Chicos*, web-site visitors) become paralyzed and rapidly begin to turn away from the problem. As audiences turn away, the media is encouraged to move on to cover other stories. Compassion fatigue sets in, and, according to Moeller, '. . . militates both against caring and against action' (Mueller, 1998, p. 53).

The case of *Por Los Chicos* can be interpreted by following the compassion fatigue framework. But there is also more to this particular case than the use of decontextualized images of the universal child and the commodification of suffering that leads to compassion fatigue. What is interesting (and disturbing) is the fact that the target audience of the *Por Los Chicos* website were for the most part other Argentines (both in Argentina and in the Global North) who supposedly could understand the Argentine context better and who could potentially relate to a more embedded subject. Faced with this situation, the NGO still attempted to construct the network of donors based on universalizing notions of responsibility and care (i.e., it is everyone's duty to help and to care simply because of our common humanity). As Silk (2004) explains, in the case of distant others, moral motivation (the grounds upon which people are induced to care for others) often identifies *universalism* as a base for action. The NGO also used the 'stranger relationship' model, that is, the model where the donor will never meet or even hear of the intended beneficiary (Silk, 2004). But as the failure of this particular network in sustaining caring over time demonstrates, universal motivations for caring and the stranger relationship model are limited because they do very little to engage distant others in any meaningful way. They do almost nothing to effect and to sustain change on the ground. They do very little to engage people beyond merely the provision of aid, and they do very little to move beyond uneven geographies of care that are characterized by unequal relationships between donors and recipients.[5]

Reflecting on the use of the dissembedded iconography of the universal child of developing nations and the type of responsibility and care it suggests is crucial because as Ruddick (2003, p. 341) indicates, 'what is at stake here is not the improved physical welfare of children, but the modes by which this is achieved'. I argue that in the case of *Por Los Chicos* the use of decontextualized images of children as the 'universal child' results in what Bordieu and Passeron (1997) calls 'symbolic violence'. This concept, as elaborated by Bordieu, refers to a type of violence that is exercised at the level of symbolisms and whose function is not that of violence per se but rather one of natural difference or distinction (Marcoulatos, 2001). As Marcoulatos (2001, p. 17) explains, violence is exercised '. . . through the unreflective imposition . . . of representative existential manners of certain dominant groups . . . its distinctive virtue is that it is not typically recognized as violence when exercised, which allows it to be more effective'. I argue that the case of *Por Los Chicos* is another example of symbolic violence towards others, in this case, children. The iconography of the universal child employed by *Por Los Chicos* acts as a gaze that turn children into what Escobar (1995, p. 155) calls a spectacle or an '. . . objectifying regime of visuality that dictates how [children] are apprehended'.

In the case of *Por Los Chicos*, however, such regime of visuality has been unsustainable. In fact, the spectacle became problematic when the *consequences* of mobilizing such discourses and images of children did not necessarily mesh with the *expectations* of those who initiated and mobilized such constructions in the first place. For example, middle and upper class Argentines just pointed and clicked on a computer screen from the comfort and safety of their own homes and felt that they had done something to help their fellow citizens in need. Hungry and needy children were not far away from them, in fact, they were outside in the streets, perhaps right outside the doors to their own homes. Instead of creating a more responsible geography of care, the model followed by the NGO ended up enlarging and creating distance (both physical and emotional) among people who were otherwise close-by and were not distant others at all. The same could be said for Argentines abroad (in Europe, in the US) and for anybody else who also bought into the idea that one click of the mouse was a step towards caring for children in need. Instead of creating a more personal and responsible connection among distant others, the model followed by *Por Los Chicos* helped consolidate both the geographic and emotional distance between more affluent Argentines in the Global North and hungry children who were living in the context of violence and poverty in the streets of most urban centers in Argentina.

Moreover, these universal constructions of hungry Argentine children soon began to circulate through other global networks of aid beyond the original one, they took on lives of their own, and became subject to appropriation by others. The result of such conflicting mobilizations and interpretations of discourses and images about hungry children was a series of trans-local conflicts that did nothing to help children in the long term but that instead, were used to further demarcate the divisions between a rich Global North and a poor Global South. Nothing illustrates this better than the surprise that local NGOs, development practitioners and activists got when they decided to accept medical donations from other networks of global donors that have been assembled as part of an agreement between the United States and the government of the city of Buenos Aires. The aid was targeted to needy children in the interior, in particular in the province of Tucumán (the place where most children's deaths had occurred, and one of the places in Argentina that was heavily covered by the global media). As containers were being opened, instead of medication what was found was medical waste (including expired medication, used dirty needles, bloody medical clothes, expired food, and even Global North patients' medical histories).[6]

Activists blamed the Argentine government and corporations in the Global North for contributing to the creation of illegal networks for the disposal of medical waste that gave hospitals and companies in the Global North an economic benefit (by cheaply disposing of medical waste in the Global South). Argentine government official denied any responsibility and one official immediately blamed it on the capitalist greed of the United States and the Global North (certainly an easy way out in Latin America). In the meantime, regular people in the street began reacting against donations from Europe and the US. The comments commonly heard were things like 'Who do they think we are, Africans? We are not *that* desperate, we are Argentines, we live in a country of abundance'. These comments unfortunately reflect people's lack of understanding of their own context in Argentina (they could not see the fact that children were actually dying of hunger in most urban centers in the country) as well as a willingness to treat an other as they have been treated. Such popular comments are further evidence of the failure of the kind of model followed by NGOs such as *Por Los Chicos*, a model that contributes to the creation of decontextualized and distanced connections even among proximate others. Anti-globalization activists also got involved, and used both the regular and alternative media to link the hunger of children in Argentina to the actions of an imperialistic Unites States, placing children at the heart of the debate on the effects of neo-liberal reform. The American ambassador to Argentina publicly denounced all this as anti-American propaganda, and a diplomatic crisis ensued, with back and forth calls between Washington and Buenos Aires. In the end, no one wanted to assume any kind of responsibility (not even at the level of diplomatic international relations), and it seemed that the assembly and organization of these now global networks of care was beyond anyone's control. In the meantime, more children died in Tucuman as a result of lack of food and medicines.

## Embodied Notions of Care and Relational Thinking

The case of the NGO *Por Los Chicos* demonstrates that when it comes to the issue of the delivery of aid for children, (mostly) good intentions are not enough. As Dichter (2003) has recently argued regarding the development industry, development assistance for over four decades has produced negligible results, despite trillions spent and, as the title of his book clearly conveys, 'Despite Good Intentions' (Dichter, 2003). The lack of sustained effectiveness of the internet-based network of sponsors and donors created by the NGO *Por Los Chicos* demonstrates the lack of actual engagement of the participants with existing realities on the ground. Even though the NGO attempted to use a new technology to engage participants, the model continued with a distant and disembodied mode of engagement that has been pervasive in development thinking and practice for decades (see Escobar, 1995). Even those people who could or should have been interested in establishing a more responsible and committed stance towards the plight of child poverty in Argentina could probably not have done it through the framework that informed the construction of *Por Los Chicos*. The problem of the model pursued by *Por Los Chicos* was not just one of network organization, that is, it was not only related to problems of organizational structure and to the logistical difficulties that often plague the distribution of assistance and development aid. It stemmed rather from the universal, disembodied, and distant notions of care and responsibility that were, perhaps unintentionally or innocently, invoked in the construction of *Por Los Chicos* in the first place. The unfortunate unintended consequences (e.g., the symbolic violence committed against children) were intimately related to the mobilization of universal notions of care and the imagery of the universal child.

If, as I argue the case of *Por Los Chicos* demonstrates, caring about others –and caring about others at a distance—is not an easy task, what are the alternatives? If the model followed by *Por Los Chicos* and many other organizations like it is not sustainable or effective, does this mean that supporting the creation of organizations and networks to help close the gap with distant others is a fruitless effort? How can we achieve more committed and effective *practices* of care at both the institutional and personal levels? Instead of taking a pessimistic stance, I want to argue that it is possible to make a case for a normative ethics of care that is also more relevant to policy and practice. I suggest that it may be useful to begin thinking differently about what it means to act responsibly towards distant others, and about what it means to create more embodied and committed ways of caring for others. The goal is to find alternative ways of caring and acting responsibly that combine pragmatically both the proximate and the distant. I argue that we can begin to approach such an alternative by drawing insights from recent work in geography (e.g., Ettlinger, 2004; Massey, 2004), post-development studies (e.g., Rao and Walton, 2004), feminist economics (Folbre, 2001, 2004) and

feminist ethics of care (Robinson, 1999) that take relational thinking and human relationships as points of departure.

For some years now, scholars working from a post-development perspective have been
arguing that the deployment of development knowledge and practice (such as global aid
for hungry and poor children) often occurs in a networked and relational fashion
(Escobar, 1995, Rao and Walton, 2004). As Escobar (1995) has argued, in the case of
development practice, multiple users move and meet in a networked space, and in the
process they share and contest development knowledge. Thus, it is not surprising that
there are often unanticipated problems and unintended consequences such as the ones
I documented in the case of *Por Los Chicos*. Distance, lack of knowledge of realities
on the ground and lack of understanding of local cultural context can make well-inten-
tioned global aid and internet-based networking initiatives less effective than initially
intended. Thus, caring about both proximate and distant others is difficult, and the first
lesson we should draw from engaging with post-development and relational thinking
about responsibility and care is that those involved in the construction of networks of
solidarity and aid should be careful about how and for whom such geographies are
assembled in the first place.

Recognizing that caring about proximate and distant others is difficult and that it
should be approached critically, however, still does not answer the more pragmatic
and normative question posed earlier, namely *how best to care and act responsibly
about others*. Feminist economist Nancy Folbre (2001, p. 232) has argued that care
should not be considered as another commodity, and that '... personal, face-to-face,
emotionally rich relationships are crucial to the delivery of high quality ... care'. I
take Folbre's personal and committed understanding of care as a point of departure to
answer the previous question. Furthermore, I argue that some conceptual clues as to
how one might begin to answer this question can be found in the feminist literature
on critical ethics of care. The work of feminist international relations scholar Fiona
Robinson (1999) is particularly inspiring. Robinson has argued that one of the problems
with common thinking about responsibility and care in the context of international
relations is that it is morally incomplete and contextually inappropriate because it
seeks to uphold impartiality by maintaining a depersonalized and distanced attitude
towards others. Robinson argues that this is also related to a dependence on a theory
of ethics that is based on concepts of universal rights and universal justice (such as
the ones mobilized by *Por Los Chicos*), concepts that in the political arena are devoid
of any connection to personal affections and interests.

As an alternative to this model, Robinson suggests a critical ethics of care that is based
on a *relational ontology* that sees people as interdependent and where the *affective* plays a
primary role. For Robinson, this approach '... starts from the premise that people live in
and perceive the world within social relationships' (Robinson, 1999, p. 2). Robinson
argues that a critical ethics of care '... recognizes that those relationships are both the
source of moral motivation and moral responsiveness, and a basis for the construction
and expression of power and knowledge' (Robinson, 1999, p. 2). Finally, Robinson
argues that the values of an ethics of care are '... centered on the maintenance and promotion of good personal and social relations among *concrete* persons, both within and
across existing communities', and that these values are '... relevant not only to small
scale or existing personal attachments but to all levels of social relations' (Robinson,
1999, p. 2).

Robinson's argument resonates with recent work by critical human geographers who,
from different perspectives and examining different issues, are working on the relations
between ethics and geography and seeking a more politically committed approach to

responsibility and care. In spatial terms, this emerging body of literature is also beginning to consider responsibility and care *relationally* rather than territorially (see, for example, Ettlinger, 2004; Massey, 2004). One of the main proponents of this approach is Doreen Massey (2004), who has asked geographers to move beyond a hegemonic geography of care and responsibility that operates with the implicit understanding that '... we care first for, and have our first responsibilities towards, those nearest in' (Massey, 2004, p. 9). Massey argues that this hegemonic geography of care is utterly territorial, and that we need to move to a notion of relational responsibility that is embodied, not restricted to the very local, and that recognizes that '... the distant is implicated in our 'here' because identities ... are forged through embodied relations which are extended geographically as well as historically' (Massey, 2004, p. 10). Massey (2004) and Robinson (1999) are asking us not to circumscribe care as a moral obligation for the private sphere (on this see also Folbre, 2004), and at the same time, not to fall into the trap of disembodied universalisms. Thinking about geographies of responsibility in such relational ways is important in the theorization of *alternative strategies of care*. In other words, the literatures on feminist ethics of care and relational responsibility in geography give us a way to think about how caring about others and acting responsibly on their behalf can be made relevant to us and to our locally-embedded experiences.

What might this entail, then, at the level of practice? What might be the lesson of applying this way of thinking for those involved in efforts to care for children and young people in need? Robinson (1999) suggests that what is at stake here is *learning to listen to and to be critically attentive and responsive to the needs and suffering of others* (as opposed to acting upon presuppositions of what ought to be done based on more abstract ideas about universal justice and rights). Robinson's suggestions also intersect with new writing in the critical development and post-development literature. This literature emphasizes listening to people's needs and harnessing people's collective aspirations for positive social transformation (see for example Appadurai, 2004, and Sen, 2004). In other words, caring about others might also involve allowing people to learn to be responsible for themselves and to care for and protect one another while at the same time acknowledging our roles and positions of power in a context of both close and distant social relationships. It is about recognizing the interdependent nature of ourselves without re-creating new dependencies; it is about striving to create equality of agency[7] without forgetting the unequal personal and structural power relations that bind us together.

It may sound difficult to take these very complex notions of responsibility, care and development thinking to the level of practice, to translate from the language of academics to the difficult and very concrete practice of helping children in need. But there are already examples of NGOs that are attempting to do exactly this. These are not ideal cases, but rather incipient attempts at acting responsibly on behalf of children in a more relational and embodied fashion and in ways that also involve other young people in the construction of their own future and communities in the process.

One such organization is the *Red Solidaria* (Solidarity Network), an NGO founded in 1995 in Argentina, that from the beginning aimed to find ways to help poor children, young people and needy families by establishing meaningful personal relations between those in need and other people and organizations who were willing and capable of helping. Over time, the organization created a large network of volunteers whose goals are to create a new culture of solidarity among Argentines, and in particular among young people. Today, there are many local networks of volunteers spread out across Argentina, each of them acting in their respective communities but also linked to each other through the larger organization. Rather than focusing on a narrow set of initiatives, the projects of the *Red Solidaria* involve a wide range of interrelated activities, from small investment and micro loan programs to a solidarity course aimed at young professionals who are interested in learning to think critically about strategies of care to even a school of future leaders and values aimed at senior year high school students with an interest in social work and development. Together with these programs the *Red Solidaria* is also involved in some more traditional programs, such as the *Programa NUTRIR* (Nutrition Program), which is a project to combat child hunger. But even this program goes beyond the mere provision of food for hungry children, and also includes an assessment of the dynamic contexts and conditions under which child malnutrition develops (even accounting for different types of mother-father-child relations and the emotional roots and consequences of malnutrition) in order to develop procedures and policies to combat it.[8]

Also important is the work that the *Red Solidaria* has been doing with young people (from 14 to late 20s). The *Red Solidaria* has long encouraged the participation of young people to find solutions to some of the more pressing social problems affecting Argentina, such as the plight of child hunger that I described earlier in this paper and the drama of missing children. In this context, young people are active participants not only in the formulation of ideas and initiatives but also in their execution. In July 2005, the network organized a gathering of young people in over 170 localities across Argentina. In each place, young people were actively debating what it means to care for others in the context of Argentina today and exchanging ideas on how best to help others (from children to young people to adults in need), including initiatives which will soon be implemented across the country.

For example, different groups of young people have conducted local surveys of needs in their respective communities, specifically targeting children and older adults living alone. Based on the results of their surveys, different groups proposed solutions that were targeted to the particular needs of each population. In one case, a local group of young people has organized a program called 'The Little School' on Sunday afternoons to provide a safe and healthy gathering place for children. The program was created after young people realized that many children were being left alone even during weekends (because of parents' work demands) or did not have access to a full meal during weekends when the public schools where poor children eat during the week were closed. The program is run exclusively by young volunteers and it includes performances, music and other forms of entertainment and artistic expression for children, created, organized, and delivered by young people themselves. Other local activities that young people have conducted included a toy repair workshop, where children and young people work together fixing toys, and ecological patrols where young people paired with an organization that helps children and young people with different disabilities and together embarked in trash removal and neighborhood improvement tasks. In all cases, young people are able to help and care for others and for their communities while developing and using their own creative capacity and skills.

The *Red Solidaria* has also taken advantage of the internet, but in a more powerful and meaningful way than organizations such as *Por Los Chicos* have done. For example, those young people who wanted to participate in the activities of the network but who were either far away or could not be physically present in one of the many locations where young people were meeting nevertheless were able to play an active role through a virtual congress, where they were able to post ideas and debate in real time with young people across the country. In this case, the world wide web was used as an avenue that allowed for ideas of distant others to be debated among local groups and possibly be executed locally, but far away from their source. This, perhaps, is an example of Massey's (2004, p. 10) notion of 'the distant implicated in our here', and one of the ways in which young people are enacting new ways of caring. The participants in the *Red Solidaria* may not be re-inventing or re-conceptualizing the concept of care, but they are at least exploring new and pragmatic ways to care for others by combining the proximate and the distant. There is a lot more that could be said about organizations such as the *Red Solidaria* in Argentina, but I hope that the brief description above is sufficient to demonstrate that the relational and embodied approach to responsibility and care that I described above can be translated to new ways of caring about others through practices that harness people's (including children and young people) creative capacities and collective aspirations.

## Conclusion

In this paper, I critically contrasted the workings of two organizations (*Por Los Chicos* and the *Red Solidaria*) that aim to help hungry and needy children in Argentina. Both organizations emerged in the context of neo-liberal reform and the shift towards a voluntary system of caring for those in need as the state abandoned its traditional role in social welfare. I have positioned the experiences of the two organizations relative to broader theoretical discussions of the relations between geography, ethics, and care. My intention was to relate some of the concerns of children's geographers with new ideas that are emerging in human geography. The paper aimed to explain how children are at the center of the inequities generated by globalization and how different emerging geographies of responsibility and care have a direct effect on children's livelihoods.

*Por Los Chicos*' main goal is to collect funds for food donations. Its distinctive approach is based on the use of the internet to connect 'responsible' or caring sponsors (mostly private corporations) with people who are casual internet users and who become donors by clicking on advertising banners appearing in the organization's website. While the approach of *Por Los Chicos* is innovative, in particular in Argentina, the organization has been unable to sustain an adequate level of interest from both donors and sponsors. As a result the total number of funds collected and aid delivered has declined since the organization was created. On the other hand, the *Red Solidaria* is a nationwide network of volunteers that specifically works to aid children in need through unique local initiatives. The *Red Solidaria* also relies on the ideas and volunteer work of young people to help children in different communities. The *Red Solidaria* has utilized new communication technologies such as the internet, but in its case the internet has been used it as a mobilizing tool for new creative ideas that can enhance the goals of the organization.

Both *Por Los Chicos* and the *Red Solidaria* have extended well beyond their initial domains, becoming national and even trans-national in scope. Yet, despite reaching people in Argentina and beyond, the approach of *Por Los Chicos* does not seem to have created a more committed or responsible ethics of care among those who participate. The organization's appeal to a totalizing notion of care and its use of universalizing representations of children has been insufficient to personally involve people with the plight of needy children. As a result, the long term effectiveness and sustainability of the organization remains in question. On the other hand, while most of the actual aid and caregiving for children that the *Red Solidaria* sponsors depends on local volunteers, the organization has been able to tap into the creative capacity of committed young people in other localities through activities like the virtual congress. The *Red Solidaria* also has national and international reach through the use of the internet, but its approach to caregiving encourages active participation and involvement. Relative to *Por Los Chicos*, the *Red Solidaria* has been better able to articulate the proximate and the distant in its attempt to provide aid for children in need because it has emphasized a more relational and personal approach to caregiving. In her compelling analysis of economies of care in western societies, Nancy Folbre (2001) describes caring labor as labor that is done on a person to person basis and for reasons that include affection and respect. It is along the lines of this definition of care that the *Red Solidaria* has been able to create a more effective geography of care than *Por Los Chicos*.

In conclusion, the comparison of the trajectories of these two organizations provides a useful platform from which to begin thinking about the challenges of creating more responsible geographies of care for children. Neither of the two organizations should be considered as either an ideal or a total failure, but there are important insights to be learned from their experiences. One of the lessons that can be learned from the case of the *Red Solidaria* is that there are certain ways of caring for others that could and should be incorporated in the thinking and practice of organizations that aim to help others—from the formulation of development plans at different scales to the more concrete issue of the construction of networks of aid and solidarity. As Robinson (1999) argued, the goal should be building long-term commitments that recognize that solutions to fundamental problems can be found at the level of relationships; ultimately social relations are the critical tool from which to begin to address our most fundamental problems. At issue, then, is *institutional change* that reflects a more committed, embodied, and relational way of care and caring for others. At a normative and pragmatic level the lesson here is that the organizations and institutions we create to help others should be built and organized based on these embodied and relational premises of care if we are to find more effective and sustainable solutions.

## Notes

1. Research on the relations between ethics and space has proved particularly fruitful in recent years. For a review of the relation between ethics and geography see Proctor (1998a,b, 1999), and Proctor and Smith (1999).
2. Capacity building in the development context is often equated with empowerment and the bottom-up process of building of human and social capital (Taylor, 2004). The digital divide refers to differential access to computer communication technologies (e.g., who is connected to the internet and who is not) and to the quantity and quality of access. Research on the relations between development, capacity building, and the digital divide typically examines the role that differential computer access play in widening social gaps throughout society, particularly among young people.
3. The website of the NGO can be visited at www.porloschicos.com
4. Data on sponsors, donations and meals is available at the 'Por Los Chicos' official web page (www.porloschicos.com. Data last accessed May, 2006).
5. An extended review of the limitations and criticisms of embracing universalizing ways of caring can be found in Silk (2004).
6. This was widely reported by the media and documented mostly by grassroots activists in the alternative media, in particular through web sites and blogs, such as www.elcacerolazoonline.com.ar, where images of the open containers with medical waste are still available to the public.
7. Equality of agency in this context refers to the idea that, besides providing access to human and physical capital, development should be concerned with facilitating people's access to cultural and social capital (Sen, 2004)
8. More information about the Red Solidaria as well as details about the many programs it sponsors can be found at its website, located at http://www.redsolidaria.org.ar

## References

Aitken, S. (2001) *Geographies of Young People: The Morally Contested Spaces of Identity,* London: Routledge.
Aitken, S. (2004) Placing children at the heart of globalization, in Barney Warf, Kathy Hansen and Don Janelle (eds) *World Minds: Geographical Perspectives on 100 Problems*, Norwell, MA: Kluwer Academic Publishers, 579–84.

Appadurai, A. (2004) The capacity to aspire: culture and the terms of recognition, in V. Rao and M. Walton (eds) *Culture and Public Action*, Stanford, CA: Stanford University Press, 59–86.

Auge, C. (2002) Argentina: Life after bankruptcy. Le Monde Diplomatique Online Edition, September 2002. Available at http://mondediplo.com/2002/09/13argentina. Last accessed October 4th, 2004.

Barnett, C. (2005) Ways of relating: hospitality and the acknowledgment of otherness, *Progress in Human Geography*, 29(1), 5–21.

Bosco, F. (1998) State society relations and national development: a comparison of Argentina and Taiwan in the 1990s, *International Journal of Urban and Regional Development*, 22(4), 623–42.

Corbridge, S. (1993) Marxisms, modernities and moralities: development praxis and the claims of distant strangers, *Environment and Planning D: Society and Space*, 11, 449–72.

Ditcher, T. (2003) *Despite Good Intentions: Why Development Assistance to the Third World Has Failed*, Amherst, MA: University of Massachusetts Press.

Dyck, I. (2005) Feminist geography, the 'everyday', and local–global relations: hidden spaces of place-making, *The Canadian Geographer*, 49, 233–43.

Ettlinger, N. (2004) Towards a critical theory of untidy geographies: the spatiality of emotions in consumption and production, *Feminist Economics,*10, 21–54.

Escobar, A. (1995) *Encountering Development: The Making and Unmaking of the Third World*, Princeton, NJ: Princeton University Press.

Folbre, N. (2001) *The Invisible Heart: Economics and Family Values*, New York: The New Press.

Folbre, N. (2004) *Family Time: The Social Organization of Care*, New York: Routledge.

Hale, B. (2002) Web Power Fights Argentine Poverty. BBC News, 7 June 2002. Available at http://news.bbc.co.uk/1/hi/business/2027508.stm. Last accessed October 4th, 2004.

Herod, A. and Wright, M. (eds) (2002) *Geographies of Power: Placing Scale*, Malden, MA: Blackwell.

Horton, J. and Kraftl, P. (2006) What else? Some more ways of thinking and doing Children's Geographies, *Children Geographies*, 4(1), 69–95.

Katz, C. (2004) *Growing Up Global: Economic Restructuring and Children's Everyday Lives*, Minneapolis, MN: University of Minnesota Press.

Keck, M. and Sikkink, K. (1998) *Activist Beyond Borders: Advocacy Networks in International Politics*, Ithaca, NY: Cornell University Press.

Marcoulatos, I. (2001) Merleau-Ponty and Bordieu on embodied significance, *Journal for the Theory of Social Behaviour*, 31, 127.

Massey, D. (2004) Geographies of responsibility, *Geografiska Annaler*, 86, 5–18.

Mitchell, K., Katz, C. and Marston, S. (2004) *Life's Work: Geographies of Social Reproduction*, Malden, MA: Blackwell.

Moeller, S. (1998) *Compassion Fatigue: How the Media Sell Disease, Famine, War, and Death*, New York: Routledge.

North, P. and Huber, U. (2004) Alternative spaces of the 'Argentinazo', *Antipode*, 36, 963–84.

Proctor, J.D. (1998a) Expanding the scope of science and ethics, *Annals of the Association of American Geographers*, 88(2), 290–6.

Proctor, J.D. (1998b) Ethics in geography: Giving moral form to the geographical imagination, *Area*, 30, 8–18.

Proctor, J.D. and Smith, D.M. (eds) (1999) *Geography and Ethics: Journeys in a Moral Terrain*, New York: Routledge.

Rao, V. and Walton, M. (eds) (2004) *Culture and Public Action*, Stanford, CA: Stanford University Press.

Ribeiro, G. (1998) Cybercultural politics: political activism at a distance in a transnational world, in S. Alvarez, E. Dagnino and A. Escobar (eds) *Cultures of Politics, Politics of Culture: Re-visioning Latin American Social Movements*, Boulder, CO: Westview Press, 325–52.

Robinson, F. (1999) *Globalizing Care: Ethics, Feminist Theory and International Relations*, Boulder, CO: Westview.

Ruddick, S. (2003) The politics of aging: globalization and the restructuring of youth and childhood, *Antipode*, 35(2), 334–62.

Rheingold, H. (1993) *The Virtual Community: Homesteading on the Electronic Frontier*, New York: Addison-Wesley.

Robins, K. and Webster, F. (1999) *Times of the Technoculture: from the Information Society to the Virtual Life*, New York: Routledge.

Salamon, L.M. and Anheier, H.K. (1999) *Global Civil Society: Dimensions of the Nonprofit Sector*, Baltimore, MD: John Hopkins Center for Civil Society Studies.

Sen, A. (2004) How does culture matter? in V. Rao and M. Walton (eds) *Culture and Public Action*, Stanford, CA: Stanford University Press, 37–58.

Silk, J. (2004) Caring at a distance: Gift Theory, aid chains, and social movements. *Social and Cultural Geography*, 5(2), 230–51.

Smith, D. (1997) Geography and ethics: a moral turn? *Progress in Human Geography*, 21(4), 583–90.

Smith, D. (1998) How far should we care? On the spatial scope of beneficence, *Progress in Human Geography*, 22(1), 15–38.

Staeheli, L. and Brown, M. (2003) Where has welfare gone? Introductory remarks on the geographies of care and welfare, *Environment and Planning A*, 35, 771–7.

Tapscott, D. (1998) *Growing Up Digital*, New York: McGraw Hill.

Taylor, D.R.F. (2004) Capacity building and geographic information technologies in African development, in Stanley Brunn, Susan Cutter and J.W. Harrington Jr (eds) *Geography and Technology*, Norwell, MA: Kluwer Academic Publishers, 521–46.

Wenar, L. (2003) What we owe to distant others, *Politics, Philosophy and Economics*, 2(3), 283–304.

# Changing Livelihoods, Changing Childhoods: Patterns of Children's Work in Rural Southern Ethiopia

TATEK ABEBE

*Department of Geography/Norwegian Centre for Child Research, Norwegian University of Science and Technology, NTNU, N-7491 Trondheim, Norway. E-mail: tatek.adebe@ svt.ntnu.no*

## Introduction

Recent works in childhood studies have focused on the westernization of global child-hoods (Stephens, 1995; Aitken, 2004). They argue that the current crisis in notions of childhood is related to profound changes in the new globalized modernity in which 'the child' was previously located. Stephens (1995) maintains that the historical construction of the world's children suggests the complex globalization of a once localized Western construction of childhood. The globalization of childhood is attributed to the politics of knowledge construction, the colonization of the global South and the imperialism of devel-opment aid, which valorize hegemonic, normative, 'modern', Western notions and under-standings (Escobar, 1997; Jones, 2004). However, recent works on children geographies have also contributed a growing body of literature that documents the living conditions of diverse childhoods in non-Western contexts, including that of orphans (Ansell and van Blerk, 2004; Robson, 2004; Ennew, 2005) and working and street children (cf. Hecht, 2000; Ansell, 2005). Findings from this literature contradict conventional and global wisdoms of childhood as being a playful, work-free, dependent, vulnerable and care-receiving phase of the life course (*ibid.*). They also challenge another crucial aspect of childhood, namely its domesticity—that is, that children are to be located inside society, a family and home, a private dwelling and a school (Ennew, 2002).

The focus of research on the place of children in social reproduction in the global South has so far been on their participation in paid and care work (Ansell and Robson 2000; Robson, 2004), productive economic activities (Bass, 2004) and within the context of HIV/AIDS epidemic (Ansell and Young 2004). In Ethiopia, studies have explored the livelihood strategies of children within the household economy in urban contexts (Aptekar and Abebe, 2001; Poluha, 2004). However, the various ways in which rural children and young people are multiply affected by, and try to negotiate, the uneven manifestation of globalization is least explored by empirical research. The central aim of this article is therefore to explore how interconnected global processes are creating disruptions in the production and rural livelihood patterns of children in families in southern Ethiopia. It specifically looks at how development strategies fuelled by globalization (e.g., unfair global trade, the HIV/AIDS epidemic and Structural Adjustment Programs) are altering children's work patterns, the nature and type of work they participate in and their social relationships within the community. The article elucidates the changing work patterns of children in the Gedeo ethnic community in which small-scale subsistence production formerly met basic household needs.

The transition in livelihoods is characterized by a move towards market-driven production to meet national priorities for economic development in an era of globalization. Historically, this change began to take shape when farmers increased production of 'cash-rich' crops, like coffee, at the expense of a subsistence crop, *enset*, a drought resistant, banana-like plant and a staple food for the population of southern Ethiopia. The expansion of coffee as a mono-culture and a dominant source of livelihoods is thus at the core of this transformation. Based on child-focused qualitative research methods, I argue that children's dynamic roles in household (re-)production should be placed within the framework of interconnected globalization processes, which not only reinforce one another, but are also fostering new forms of socio-structural inequality. I shall do this by offering a nuanced account of children's everyday survival strategies, as these are an integral part of the wider changes taking place in family and community livelihood strategies.

## Methods

The empirical material for this study is drawn from successive fieldwork trips conducted over a period of seven months[1] in southern Ethiopia. These involved observation, conversations, focus-group discussions and in-depth interviews with children, young people, community members and development workers. These methods were accompanied by a number of task-oriented child-focused activities, such as writing essays in a school (where I taught voluntarily) and encouraging children to participate in story writing, sport and music competitions. These approaches enabled me to generate in-depth knowledge on how 'external' processes are experienced by and manifest themselves in children's everyday lives. They also helped me establish trust and confidence between myself as an Ethiopian researcher and the children as participants in society. The methodologies I employed and the multiple positions I held during the research process are based on approaches which have successfully been used by other researchers (Woodhead, 1999; Punch, 2001).

The information obtained from the various sources has been synthesized along with data from the field logs and semi-participant observation. Moreover, the fieldwork has also been informed by my previous professional engagement and research experience with orphans and working children in the region (Abebe, 2002). The data presented in this paper—which involves a narrated contextualization of children's work experiences—looks closely at the age, gender, type and nature of work, work spaces, work cycles and spatiality of livelihoods of children who participate in diverse reproductive and productive activities, including coffee production, Ethiopia's single major export product. The article is structured as follows. The first section gives an account of the context of the study and discusses conceptual and theoretical debates pertaining to children, livelihoods and social reproduction. Next, I explore the influence of transformations of family livelihood strategies on children's participation in work in time and place and within the contexts of everyday life. Finally, I discuss the implications of the changing work patterns of children in the light of far-reaching policy issues and economic and political structures.

## Gedeo in the Global Context

The Gedeo district is one of twelve districts which lie within Ethiopia's Southern Regional State. It has one of the highest population pressures, with an average rural density exceeding 400 persons per square kilometer (CSA, 2000). Unlike the grain-producing region of Ethiopia's north, it is inhabited by subsistence agricultural farmers who, for generations, depended on a drought-resistant root crop, called *enset*, as their staple food. According to Hamer (1987), income from commercial crops simply supplemented the limited need for cash. Like many other *enset* areas in Ethiopia which have a high population-carrying capacity (Bevan and Pankhrust, 1996) Gedeo had been relatively self-sufficient, and the people had maintained stable rural livelihoods.

In the past couple of decades, however, there has been a growing emphasis on the production of cash crops based on the ideal that the expansion of exportable products and competition in the global market is necessary for rapid economic development. Given the growing significance of the cash economy, which roughly coincided with the exponential rise in the price of coffee on the world market, farmers in Gedeo believed that they could buy everything they needed, including *enset*, which grows alongside coffee trees. *Enset* production was subsequently marginalized, even though its shade provides coffee beans with the most conducive environment for their gradual ripening (Tadesse, 2002). At the same time, farmers who formerly produced multi-purpose crops were tactically made to concentrate on the production of coffee alone. This was made apparent when, in the late 1970s, the socialist regime institutionalized and implemented a system of farmers' cooperatives in which local farmers were supplied with agricultural inputs that promoted coffee production. Today, farmers sell their produces to cooperative shops, locally known as coffee unions, which set the farm-gate price for fresh coffee, and collect and transfer it to the central market in Addis Ababa. It is in the capital city that the coffee is made ready for export to the global market.

Presently, rural livelihoods in Gedeo are based mainly on the production and sale of commercial crops like coffee, *chat* (a mild stimulant leaf) and fruits. The growing importance of coffee as a single source of cash over the self-reliant subsistence economy has had a considerable impact on farmers' livelihoods and living conditions. Over the period from 1990 to 2002, the wealth of farmers has been severely eroded due to a continuous fall in world coffee prices (UNDP, 2005). As a result, farmers who had used most of their land holdings for coffee production eventually fell into the trap of livelihood insecurity when the income from it ceased to be reliable. With plummeting coffee prices, farmers began to migrate to the towns and cities in search of alternative livelihoods. Problems were further compounded by a rapid growth in population and a shortage of farm land, as more and more children came of age and entered into adulthood. Seasonal and permanent out-migration to obtain off-farm employment became one of several livelihood strategies for farmers in southern Ethiopia. However, adult labor migration profoundly reshaped the livelihood strategies of children and women who are left behind locally. Moreover, given the widespread culture of polygamy and absence of safe sex, it has fuelled high incidences of HIV infection and AIDS-related mortality in Gedeo, despite the fact that average infection rates from HIV in most parts of rural Ethiopia is relatively low (CSA, 2000). Consequently, and although official figures are difficult to come by, the Chichu Peasant Association in the Gedeo district from which the data for this study are drawn is home to a large number of children who have become orphans due to HIV/AIDS.

Furthermore, studies indicate that dramatic shocks in livelihoods associated with an insecure income are the main reasons for the plight of farmers' poverty and food insecurity in southern Ethiopia (OXFAM, 2002). With near-perfect conditions, farmers in Gedeo district produce one of the finest crops of sun-dried, organic Arabica beans in the world, which grow in almost every spot of the lush, fertile, green countryside. However, the world coffee market does not distinguish organic beans produced by peasant farmers from low-quality, mixed blended coffee produced using modern technology and inputs, such as chemical fertilizers and steam washing. In addition, global restructuring of the coffee market from one which previously regulated supply and price to a more liberalized approach[2] in which the market fixes both demand and price is at the root of this devastating situation, which is greatly complicating the lives of small-scale coffee-producers.

Demelash[3] is a coffee-producer in Gedeo. He works very hard on his 0.75-hectare farm with his wife and six children. He has also taken on the responsibility for his sister's two children after she died of AIDS. Demelash complains that farmers are not receiving the price they used to obtain for their produce:

> Last year I did not even go to pick the berries because the income barely covers the labor of the children engaged in the production ... let alone what is needed for their nutrition, schooling and medication. As a result, I am cutting back on food and am not able to pay school fees, nor buy exercise books and uniforms for my school-age children. The coffee crisis has undermined the livelihoods of many farmers in Gedeo. In some cases, [impoverished] farmers have ripped up the thriving bushes laden with beans to plant maize, simply to survive. I myself have taken to sell the corrugated tin from my roofs, which once set me apart from the poorer neighbors. Why should I grow coffee when all it does is ruin my life?

The imbalance in international trade terms, which is dictated by powerful countries and economies, is a sharp illustration of the challenges involving many commodities on which developing countries rely heavily. In Ethiopia, close to a million households depend directly on the coffee sector for their entire income, and a further fifteen million people derive their livelihoods indirectly from its sales (CSA, 2003). The importance of coffee in securing vital foreign currency cannot be overestimated because it accounts for over 60% of Ethiopia's foreign exchange revenues (UNDP, 2005). However, in the last few years this crucial income has declined by more than half[4] (OXFAM, 2002), despite an overwhelming increase by over two-thirds in the amount of coffee exported since the mid-1990s (UNDP, 2005). As a result, Ethiopia's economy is teetering on the brink of collapse, which poses a major hurdle to its Poverty Reduction Strategy Program—a successor to a failed Structural Adjustment Program (SAP) lasting over 10 years—whose central plank is economic development led by agriculture, the only area in which it has an obvious 'comparative advantage'. Unfair global trade also promises to throw off course the World Bank and International Monetary Fund (IMF) debt-relief formula, which is otherwise dependent on sustainable strong growth in exportable agricultural products such as coffee.

Apart from falling price of coffee, the impacts of macro-economic, structural changes associated with the SAP are acute, particularly in rural areas which received very little of the initial investment in health and education. For farmers like Demelash, the SAP meant a rise in land tax and the removal of subsidies from agricultural inputs, which became a major disincentive for the production of commercial crops that were excessively promoted at the expense of local subsistence economies:

> If I want to re-plant *enset* on my coffee farm plot, I will have to wait for a minimum of four years to harvest the yield for consumption. In addition, because I am indebted to the cooperatives who give me loans to buy fertilizers, I am obliged to sell the coffee beans to them at a fixed price. If I can save it for later, I know I could make a little more money [...] it is a whole circle of poverty.

The restructuring of Ethiopia's economy by the SAP has conditioned childhoods and exacerbated the marginalization of children. The slashing of public expenditure placed the burden of responsibility on hard-earned private resources. And those who least can afford it (i.e., women, children, the poor and the working class) suffer from it in multiple ways (Aitken, 2001). The government's withdrawal of social spending produced new pressures on children, particularly for those from the poorest group, who became contributors to family income and, in the worst case, had to provide entirely for their own welfare (Ennew, 2005). Consequently, the burdens of social reproduction fell forcefully on children and young people, who began to shoulder most of the household responsibilities at the expense of, among other things, their schooling. Thus, the SAP can be identified as a contributing factor in reshaping the local reproduction patterns of families and communities, with a direct impact on childhood deprivation, child labor, mortality, morbidity and illiteracy. Moreover, it can be shown that the SAP has exacerbated orphanhood due to the spread of the HIV/AIDS epidemic (Ansell and Robson, 2000), which, given the contexts of poverty, generates and amplifies the very conditions that enable it to thrive.

### Children, Livelihoods and Social Reproduction

Theoretical concepts that are generally used to understand women's roles in social reproduction can also be applied to understanding children's roles in social reproduction (Bass, 2004). For instance, contemporary frameworks used to explain the undervalued labor of women in relation to men (i.e., the gender hierarchy) are analogous to the undervalued labor of children in relation to adults (i.e., a socially constructed age hierarchy) (*ibid.*). Children in Ethiopia perform numerous domestic and productive activities as a contribution to the survival strategies of the households in which they form an active part (Admassie, 2003). In particular, their household (re-)production roles is a structural necessity for the maintenance of most rural households. Children's help with domestic chores is highly valued, and parents consider it children's social responsibility to contribute to the well-being of their families (Verhoef, 2005). However, as households undergo restructuring in response to economic changes, children's participation in work is not only becoming indispensable, it is also assuming new and dynamic trends. Moreover, their participation has increasingly begun to be carried out independently of adults' involvement or as an extension of women's work, which, in most instances, is undocumented and unaccounted for in economic statistics (Bass, 2004). Placing children's work within the frameworks of age and gender alone undermines the significance of historical and contemporary structural forces, which not only widen existing hierarchies in class, ethnicity, religion, title, age, and gender, but also create new socio-structural differentiations and inequalities.

Social reproduction encompasses a broad range of practices and social relations that maintain and reproduce particular relations of production along with the material and social grounds in which they take place (Katz, 2004). Katz points out that, among other things, social reproduction entails the reproduction of the population and the means by which people produce their subsistence. It encompasses daily and long-term reproduction of both the means of production and the labor power whereby people work. Social reproduction hinges on both the biological reproduction of the labor force and the physical conditions that makes ongoing production possible (Katz, 2004). In general, social reproduction includes a range of social practices that structure livelihoods and what is considered productive in a given society over time. 'It embodies the whole jumble of cultural forms and practices that constitute and create everyday life and the meanings by which people understand themselves in the world' (*ibid.*, p. x) It is, therefore, about how the relations of production are both made and naturalized through an amalgam of material social practices associated with the household, the state, civil society, the market, and the workplace (*ibid.*).

Children's productive and domestic works within the household constitute the core of social reproduction in rural areas. As a result, children's labor is not only vital in economic terms; it also comes to the fore in the continuation of societal systems themselves. However, their participation in the labor force and meaningful monetary contributions are often overlooked and occupy only marginal positions in research and public discourses. Indeed the dilemma over whether children should work or go to school, which types of work are culturally acceptable and what actually constitutes child labor is a much contested issue in academic arenas (Lange, 2000; Rita, 2001; Ansell, 2005; Nieuwenhuys, 2005). This is due to the way a given society and its structure relate to children both historically and at the present time. This section shall briefly consider whether the cultural and social influences of what Ali Muzrui (in Bass, 2004, pp. 16–36) calls Africa's triple heritage—the indigenous, Islamic, and colonial factors, which together can explain the historical roots of child labor in sub-Saharan Africa—can inform our understanding of children's work in the (re)production of everyday life in Gedeo.

The *indigenous perspective* on children's work maintains that children's work in families is part of household production and as an ongoing process of vocational education and socialization. Children are expected to contribute while at the same time learning the necessary skills that will enable them be active members of their community. It further argues that children's participation in work is vital to maintain subsistence economies and ensure the continuity of certain cultural skills (Admassie, 2003). The *Islamic perspective*, on the other hand, focuses on the role of children in the maintenance of livelihoods in a context in which women, for example, are secluded from public spaces on religious grounds. Research show how children are used as intermediaries by Muslim women, who are secluded due to the Islamic practices of purdah, to participate in trade activities and meet household economic needs (Robson, 2003). The Islamic perspective also presents child labor as a service in exchange for Quranic education (Bass, 2004). This is also the case in some Christian societies (as in Ethiopia) where children contribute their labor in order to receive a church education from religious leaders. In a historical study on the nature of child labor in the former Rhodesia, now Zimbabwe, Grier (1994) employs a *colonial perspective*. Throughout the colonial period, white colonizers and employers looked to the state for help in order to gain access to children's labor, as well as to control and discipline them. This was also the case in Togo, where girls and boys worked alongside adult family members on white-owned commercial farms (Lange, 2000). Missionaries also used the labor of African children, often as domestics in their own households and as unpaid worker-pupils on the mission-owned commercial farms (Grier, 1994). Popular with employers, African children were a cheap source of labor which raised the profit margins of many white mine-owners and of nearly all white farmers (*ibid.*).

These three perspectives, though very useful in understanding the continuity of children's (re)-productive activities, tend to overlook contemporary structural forces that disrupt the livelihoods of families. They downplay the specific context of societies, and even of the antiquity of Christianity in Ethiopia which has little to do with colonialism and the 'modern time'. Moreover, they also fail to provide sufficient insights into children's changing work patterns which are framed within the context of deeply unequal relations of power and reciprocity (Nieuwenhuys, 2005). I therefore suggest that explanations be sought that are grounded in transformation of livelihoods due to 'globalization's pandemic'[5] and trade liberalization. Put differently, the change in livelihoods which is induced by neo-liberal economic system is at the heart of the change in children's work experiences that constitutes an important aspect of contemporary childhood in Gedeo. The disintegrative impact of various strands of globalization on children's well-being, work and everyday lives has been documented in studies in the Sudan (Katz, 2004) and Zimbabwe (Robson, 2004). Pressured by the need for survival in the context of a market-driven economy, children are engaged in exploitative paid work, while capitalism is transforming the conditions of their work in ways that go beyond the imperatives of apprenticeship and socialization. Apart from the low wages and working situations of children, which are often a subject of international concern, a subtle and less noticed impact of unfair trade in coffee-reliant households is the disruptions caused in patterns of social reproduction. When families live in economically precarious situations, adults must engage in alternative livelihood strategies, partly by transferring the burden of domestic work to children. Moreover, the sustainability of household livelihoods (asset and resources) is dependent on the labor and money children contribute by engaging in income-generating activities. In what follows, I examine the participation of children in these activities from a time perspective (cycle of one year) and explore the changes in age- and gender-based forms of the division of labor that have come about as a consequence.

## The Spatiality and Temporality of Children's Livelihoods

### *The Temporality of Livelihoods: Children's Work Cycle*

Agricultural activities in any given year in Gedeo are divided into four seasons, on the basis of which the farmer's work calendar is programmed. This same work calendar also structures children's participation in work, school, play, and other social interactions. For example, during *Bonoo*, which runs from mid-August to mid-January, farmers and the members of their households are engaged in the preparation of fields and the weeding and harvesting of different types of agricultural produce. This season predominantly relates to coffee-production tasks, including picking coffee beans, washing, drying, and sorting the damaged beans from the clean ones, as well as the larger from the smaller, and transporting the beans to markets, all of which demand the intensive labor participation of children. Since the coffee beans mature irregularly on the trees, harvesting is carried out by hand in two phases of a continuum. In the first phase, called *tisha*, ripened (deep-red) berries are selectively picked, leaving the greenish ones behind to mature. This requires concentration and time, and is usually left to children and women, who are considered to be more adept at it than adult men.

The second phase of the harvest, called *bunin qocca*, requires a greater organization of household labor and the pooling of labor with neighbors and/or relatives (Tadesse, 2002). Because this phase of the coffee harvest is squeezed into a short period, from mid-December to mid-January, employing off-farm labor, particularly that of children, is a customary procedure. This practice is important because loss of yield may easily result, as coffee berries dry on the trees if they are left for a long period of time (*ibid.*). As soon as the berries are dried on the tree, the coffee tree starts to flower. Picking coffee berries once this has happened means removing the flowers, as these and the dried berries are very close to each other on coffee twigs. Children's participation in coffee picking therefore increases dramatically since careful work on the coffee treetops is needed in this short yet critical stage, which also determines future coffee yields. Intensive participation in coffee harvesting is also needed in order to minimize the danger of theft and sudden hailstorms, which could jeopardize the main source of income for most households.

The labor-intensive nature of most agricultural tasks has a direct bearing on children's participation in schools. Most children combine lessons with work, because the school they go to functions on a half-day basis. They negotiate the fulfillment economic responsibilities at home with school schedule through systematic and structured participation in work in the rest of the day, at weekends and in the summer holidays. According to the Bureau of Education[6] (2003), nearly half (46%) of the school-age children in Gedeo do not attend school. In addition, school drop-out rates are very high during coffee-harvesting seasons. Sisay (15-year-old boy) who is a class monitor in a 4th grade clearly states the problem:

> When it is the peak season for the coffee harvest, very few of us are able to come to school. Classrooms may have no fewer than twenty students [a fifth of the normal class size] because it is difficult to combine paid work with school attendance. [In some cases], we miss out so many lessons and are not sufficiently prepared to pass the first semester examinations.

Gleaning the coffee beans after a normal harvest is usually done by girls and women in the household or other poor people from the neighborhood. Gedeo tradition allows people access to private property, including coffee farms, after the owners of the farm have finished the main harvest (Tadesse, 2002). Thus, deprived farmers, women and children can enter and glean the remaining coffee berries to raise some extra income for themselves. The collected coffee beans are dried in the sun on bamboo mats stretched across beds no less than half a meter to one meter off the ground. The drying beds are usually prepared by girls. Dried coffee beans are stored in large bamboo baskets that are placed on racks 30–50 cm above the ground, to ensure proper ventilation and maintain good quality until the beans are sold. Sun-dried coffee can be stored for more than a year without risk of spoilage, allowing relatively well-off farmers to store reserves of coffee until its price starts rising (*ibid.*). However, the coffee grown by most farmers is sold immediately to washing factories (coffee unions). The government sets a bottom price for fresh coffee, although over the past 10 years the price has not risen above five Ethiopian Birr (45 US cents) (OXFAM, 2002). In this way, the expansion of cash-cropping not only affects small-scale subsistence production but also indirectly increases the state's interest in and control over the livelihood of peasants.

Apart from the coffee harvest season, children's work is also vital at other periods of the year. This includes the preparation of seedlings in nurseries and *enset* plantations, which usually occurs in March or April. Sisay describes his agricultural knowledge and the work he performs in his family as follows:

> To plant *enset*, my father and elder brother dig holes with digging sticks and drop a seedling in each hole whereas I follow them by [laboriously bending over and] scooping soil over the roots and around the lower stems of the plant. Often maize seeds are intermixed with *enset* plants because the more rapid growth of the former will shade the latter from withering.

Other activities that children are engaged in include the harvesting and sale of pineapples (January to March), the cultivation and sale of *chat* (all year round, but available to harvest only after three years), and nuts and maize (which are sown in March and harvested in July). Moreover, in the mid- and lowlands of Gedeo, cane-sugar constitutes an important source of cash for families. Most of the produce which brings in small amounts of cash and is meant for local consumption is marketed by children, particularly boys. In addition, as many of the children whom I interviewed are affected by poverty, they also actively participate in paid jobs in nearby towns to contribute whatever they can. Most children spend a lot of time performing various jobs without thinking of it as work. This is because they consider work as part of their everyday lives and cultural obligations, as well as being intricately interwoven with play activities (Poluha, 2004). The point is that children's work is structured and restructured as a result of the temporality of income secured from commercial crops. Children must also adapt to the seasonal nature of their livelihoods by engaging in other income-generating activities when agricultural activities are restricted. These activities, which I shall not discuss in detail for obvious reasons of space, are generally found in the flexible, informal economic sector, in which children find it relatively easily to integrate themselves.

### Shifting Age-based Division of Labor

Ideally, labor in Gedeo is organized in accordance with the principle of seniority (i.e., age), in which adults direct children and young people in the production processes. This is not, however, an exploitative form of authority, as Hamer (1987) argues, but one of mutual indebtedness, since adults are obliged to support the interests of their children, just as the latter are required to give their labor to increase the productivity of the gardens and herds of the former. During the coffee-harvesting season, children may spend up to six to seven hours a day picking beans for an average income of 3–4 birr (35–40 US cents). Although children are aware that they are paid less than adults who perform the same job, they consider working to be an opportunity to contribute to the household's livelihoods. That children receive less pay than adults can be explained by the system of seniority or age-based hierarchy. The role of seniority in shaping the world of children's work, play and everyday life is significant, and has been well documented in a handful of culture-specific studies (Nieuwenhuys, 1994; Hollos, 2002), including in Ethiopia (Poluha, 2004). This approach considers children's subordination in society to be grounded in a patriarchal system of relationships whereby being an elder is vested with the power to exercise authority and control over children. Consequently, children are not considered as workers in their own right (Woodhead, 1999), and what they actually do is submerged in the low-status (inferior) realm of domesticity, to which women, for example, are also relegated (Nieuwenhuys, 2000).

The generational stratification and age-grade cohort as a structure of patriarchy has, however, altered with the growing significance of the cash economy. Monetary remuneration from paid jobs provides children and young people with opportunities to exercise their own agency in how and when to spend the money they earn. This is unlike the experience of other children in most parts of rural Ethiopia, where access to productive resources in terms of cash is restricted to adults, mostly men. As Hollos observes in study of the changes in the cultural construction of childhood among the Pare of northern Tanzania,

> [T]he seniority system [in Gedeo] is undermined by the fact that most sons today are better educated and earn more than their fathers. As the survival of the family hinges more on their contribution than on land provided [to] them by the patrikin the elders control over them have diminished (2002, p. 172).

The monetization of a once subsistence economy is thus eroding an age-based seniority that forms the core of patriarchy. Concomitantly, fathers and elderly kinsmen seem to have less influence in making decisions about their children's everyday life. The shortfall of monetisation is that it increases private interests which go against culturally sanctioned traditional obligations and obedience to seniors, which are seen as the basis for a 'good childhood'. Moreover, shifts in patriarchal relations are illustrated by the pressure placed on children (particularly boys) to migrate to urban areas to seek gainful employment or participate in wage-labor within the informal sector (Ansell and van Blerk, 2004). However, by turning the viable source of livelihoods into cash, one of the stalwarts of male control; the activity of commercial cropping has also encouraged a further entrenchment of patriarchy that is rooted in gender inequalities (Simmons, 1997).

### Negotiating Gendered Boundaries of Work

Children's work is also characterized by a gender-based division of labor. In Gedeo, girls perform a full range of diverse domestic chores, which include preparing food, making coffee, cooking meals, sweeping floors, washing dishes and clothes, fetching water, minding siblings, gathering firewood, buying consumer goods from the market, and preparing *enset*, which culturally is considered to be a 'woman's crop'. The experience of Shitaye (girl, 15 years) whom I interviewed when she was doing a nursing work for her sister's child suggests how members of a household negotiate their domestic work responsibilities:

> I came here a few months ago to assist my sister who works for the coffee washing factory. I keep an eye on her child while tending some domestic chores, wash clothes and prepare food.

In response to my question about the gendered dimension of participation in work, Shitaye clarified:

> I also help my grand mother [who lives close by] I fetch her water from the communal pipe, and assist her in the kitchen when she prepares *enset* to the family ... She tells me about how important it is to learn to cook because in Gedeo a girl's decency is valued by the extent to which she can make the best meal out of this root crop. ... [The boys] perform out-of-home jobs: they work on fields or tend herds, and sell proceeds from the farms. You know they have more freedom, and are able to earn money. They can also play.

Children's division of labor in household production and reproduction is, in part, a reflection of how the adult world is structured in terms of participation and decision-making in the public and domestic spheres. According to Gedeo tradition, domestic jobs are generally assigned to women, who, at the same time, are responsible for managing the everyday household welfare and economics. Girls learn various domestic chores from the extended family members as early as the age of six or seven years as part of a continuous socialization process towards womanhood (Figure 1). However, with the growing significance of a cash economy, children's participation in work activities seems to be undergoing a gradual change. Like Shitaye, children give adults the opportunity to pursue their specialist activities, to attend market or do more arduous chores (Hollos, 2002). In addition, children independently participate in paid work by picking coffee berries, carrying them to cooperative centers, where they also work in the washing and drying process. As the cash economy is transforming the household and play activities of children into salaried work, it has led children to become 'competitive' with adults in monotonous, brutalizing systems of exploitation (Katz, 2004). In this way, children's labor is not only incorporated into the global economy but is also simultaneously marginalized, because they receive meager monetary returns from the coffee market chain. The value of the coffee they help produce in the production process declines substantially in this chain, as the beans go through various stages of processing—washing factories, cooperative shops, collection sites, and the central market—until they are exported to the coffee-roasting companies. Third-world farmers receive a paltry 1% of a final cup of the value of coffee consumed, whereas wealthy coffee companies make a third of their annual profit from Africa alone (UNDP, 2005).

Although boys and girls perform gardening and weeding jobs, generally speaking fewer numbers of girls than boys work as farmhands and in the coffee washing industries. This is mainly because girls shoulder more domestic responsibilities, which, in turn, frees boys and adults to participate in supplementary income-generating activities (Hollos, 2002). Conversely, more girls assist their mothers, who are employed in coffee-processing cooperatives in sorting, washing and drying coffee beans. In general, access to and participation in activities which generate material resources is highly gendered among the Gedeo. As a result, the former division of labor which was based on subsistence production is

**Figure 1.** Girls' reproductive work—hair making and child care (Tatek, A.)

accelerated, with boys increasingly engaged in the production and marketing of cash crops, while girls are left with the responsibility for domestic reproduction. Such a shift in the gender-based division of labor tends not only to have undesirable outcomes in terms of children's welfare but also exacerbate 'gender differences' in opportunities and exposure to different life chances. For instance, boys tend to be in a better position to contribute towards their clothing and educational expenditure and were observed to be in possession of more consumer items (e.g., radios and bicycles) when compared with girls, who are expected to put the interests of their household above their own (cf. Nieuwenhuys, 1994).

Likewise, families in Gedeo prioritize sending boys to schools because of the unequal 'cultural value' they have for children. A male child participating in school is seen as needing time to do homework, which to parents may seem to be of secondary importance for a girl attending school. As a consequence, children's work spaces and degree of mobility, both spatially and occupationally, varies a great deal by gender and age. Older children are more autonomous when it comes to making decisions about their lives than younger children of both genders. Girls' work spaces are in the invisible domestic sphere (Admassie, 2003) while boys' work places might include public spaces like bus stations, markets, streets and highways, where they trade sweet potatoes, sugar-cane, papaya, pineapple, banana and mango, as well as cooked foods like bread, biscuits, etc. In fact, boys provide a valuable link between households and the cash economy through their marketing activities. A considerable number of the child informants participating in my study are engaged in small-scale trading activities of various kinds. In Gedeo, skills in 'agricultural entrepreneurship' play a pivotal role in the economic security of children and youth. Children are thought to combine working on farms while at the same time being engaged in off-farm activities like petty trading and paid jobs to raise an income to meet the increasing demands for cash.

Gendered participation in domestic and productive work in Gedeo can sometimes be blurred and elusive depending on a number of cultural, social and economic family circumstances. For instance, both boys and girls can work equally alongside their parents of both genders assisting in the cultivation, planting, weeding, and harvesting of agricultural produce. In Gedeo, boys are culturally endowed with the right (and obligation) to be the lifeline of their households in situations when the usual head of the household (father/ husband) is lacking. However in households where there are no girls or where girls are younger, boys can assist their mothers in cooking by collecting firewood and water and going to flour mills, activities that would otherwise be delegated to a girl if there were one in the family. As a result, the birth order of children and the economic circumstances of their families exert a strong influence on the type, nature and duration of work that children participate in, where they perform that work, and their social well-being (see also Nieuwenhuys, 1994; Bass, 2004).

## Beyond the Working Child: Orphanhood and the Inter-generational Contract

Contemporary orphanhood in Gedeo is illustrative of how children assume multiple and changing roles in order to cope with the death of a parent(s) due to HIV/AIDS. Consider the life of Haimanot who lives in the suburb peasant association of Chichu along the main road to Moyale, a town on the border with Kenya. She is 14 years old and is currently caring for her youngest brother, Tesfa, 10 years old, and her ailing mother, who is HIV-positive. Her father, who worked as a day laborer in a small carpentry firm, died of AIDS three years ago. It was then when she saw for the last time two of her older brothers, who, she believes, now live in Awassa, the regional capital of the Southern Region. I met Haimanot by chance at a local NGO I often visited in order to meet orphans and carry out interviews. She wanted to apply for financial assistance to buy medicine which a doctor had prescribed for her mother. She did not know whom to ask, nor how to go about doing so. Nor did she know what to do in the event of her mother's death, for this is not a question of if, but when. She described her experience of orphanhood as involving complex feelings of ambivalence, insecurity and vulnerability:

> If a child is without parents and no one is by the side [. . .] other children and people can easily attack and humiliate. When my father was alive I use to get enough food to eat and clothing for New Year and when school opens. Having a mother makes me proud, but what would I do if she dies?

Before her mother was admitted to hospital, Haimanot worked with her selling commodities in a daily market:

> After we return home from the 'night market', I always made dinner and served coffee to my mother along with the neighbors who come by. Now we are not very close to them because we had to move out of the house due to lack of money to pay the rent.

At present, Haimanot lives in a rudimentary shack of which the roof is made partly of wood and thatch grass, covered with plastic. 'It is very difficult when it rains, and during the colder nights', said Haimanot, as we entered a slum adjacent to a churchyard, where destitute people, the elderly, and needy women and children—including herself— live so as to receive alms from people. Tesfa was boiling sweet potatoes while simultaneously sweeping the floor of their small house of approximately 10 square meters. Like many other girls in the neighborhood, Haimanot does not attend school because her mother is extremely poor and cannot afford the costs. Tesfa also dropped out of his first grade class for about a month because he was sick from malaria, one of the major causes of child mortality in the hot and humid environment of Gedeo.

Epidemiological concerns prompted by the HIV/AIDS epidemic are negatively transforming the lives of millions of children. At present, there are close to five million orphans in Ethiopia who have lost their parents due to famine, malaria, disease, war, and the HIV/AIDS epidemic — a catastrophe which the government describes as 'tearing apart the social fabric of childhood' (Abebe, 2005). Catering for the basic needs of these children requires US\$ 115 million a month, in a country whose annual health budget is only US\$ 140 million (WHO, 2000). The adverse impact of the HIV/AIDS epidemic in altering long-established inter-generational relations is immense. The epidemic has both transformed and fostered new forms of relationships between children, parents, grandparents, and members of the extended family in a number of ways. It changes the way in which children relate to their parents during the latter's long period of illness. This is very evident considering the time-warp nature of present-day orphanhood due to AIDS, in which children actually begin experiencing it long before the actual deaths of their parent(s) (Abebe, 2005). The death of adults also disrupts household systems of production and reproduction, with devastating consequences for the children left behind. Moreover, the time-lag between HIV infection and AIDS-induced mortality multiples children's workloads because parents, who are in their prime age yet terminally sick, become dependent on 'young carers' for care and the fulfillment of everyday, basic needs (Robson *et al.*, 2006).

Unlike Western ideals of childhood, in which the concern for children is predominantly emotional, children in Gedeo are sources of economic as well as emotional security. They are considered to be assets to their parents and the community both culturally and economically. As a result, investment in their childhood is rationalized for their contribution in terms of income and labor. Notwithstanding the changes in this notion due to complex socio-cultural, economic and political transformations, children continue to represent old-age economic and emotional security in societies like the Gedeo where there are no extensive social welfare programs. With the advent of HIV/AIDS, however, this 'inter-generational bargain' is becoming fundamentally transformed. Due to poverty and impoverishment, more and more young children are assuming a number of social and cultural roles that were formerly reserved for and performed by adults. The lives of Debela (13 years) and Desta (16 years) who run a small tea shop along the roadside of their grandparents' compound following the death of their biological parents is a reflection of this reality:

> We had a land which we inherited from our deceased parents, but it is far in the countryside. We are scared to live there alone, so we decided to hire someone to tend it on a share-cropping contract basis. We came here to do something which can bring us more income. I opened this shop but my brother has nothing to work. He sometimes sells banana, and brings bread and biscuits from town which I retail with tea (Desta).

> Tatek: How is the tea shop functioning?

> Desta: The problem is we lack capital. Because we don't have enough subsistence to live by now, we consume all the profit we get every day. How can we make our business grow when we have these financial constraints?

Orphans participate in society not merely as children, but also as producers, carers, entrepreneurs and decision-makers (Ennew, 2000). As evidenced by their everyday lives, they perform these roles in remarkably unique and complex contexts of poverty and marginalization. Indeed, orphans resemble other children in most aspects of their everyday lives. However, since they are living without the emotional and economic support of parents, their childhood is compromised in a number of ways. Their experiences as heads of households in the absence of adults and to maintain their well-being and livelihoods by engaging in visible, invisible, domestic and income-generating activities suggests the formation of new dimensions of childhood, gender, and relational patterns, a fact which Robson describes is 'a reversal of conventional wisdom on childhood whereby children are simply dependent on adults (Robson, 2004, p. 6; see also Guest, 2003; Nyambedha *et al.*, 2003).

In Gedeo, the onset of orphanhood due to HIV/AIDS has replaced the traditional socializing roles of grandparents and other members of the extended family by a state of 'permanency' in care-giving, despite the economic difficulties these households face. The lives of orphans like Haimanot, who does not have nor seems to be receiving any social and material support from relatives, is full of uncertainty. Such children may lack responsible role models, yet they maintain so-called 'orphan-headed' households. Although most of these households are transitory and fluid, their presence as an important feature of Ethiopian society is increasing due to the steady rise in AIDS-related adult mortality. By living outside families and adult care, orphans are undermining the general consensus that a 'family' defined by hier(patri-)archical relations across generations is necessary (Ennew, 2002, 2005). Similarly, orphans who live with extended families and/or grandparent-headed households in communities affected by the AIDS epidemic are contradicting what Burman (1995) describes as a historically celebrated correspondence between the birth of the notion of the modern nation state and the emergence of a modern or western notion of childhood. These households question the necessities of patriarchy and the social construction of nation states, which, in modern and post modern societies, are increasingly based on images of the nuclear family unit (Ennew, 2002).

## Conclusion: Some Ways Forward

This article has sought to make a contribution to the small but growing body of scholarship that is looking closely at the place of children in household (re-)production and the dynamics behind their changing work patterns. Using cases from Gedeo, Ethiopia, I reiterate my argument that children's living conditions should be set within the context of multiple globalization processes with direct local manifestations. These processes are parallel and include new forms of international trade, the impacts of HIV/AIDS and capitalism (exemplified by macro-economic adjustment programs undertaken at the expense of poor peoples' livelihoods). As these processes are occurring at the same time as the world economy is becoming increasingly liberalized, their deep impacts in destabilizing household systems of (re-)production and in depressing the material and living standards of children has been and continues to be enormous.

I argue that present-day globalization will be neither pro-poor nor pro-children as long as its outcomes continue to intensify poverty, increase dependency and perpetuate the vicious circles of indebtedness and international aid. Despite the rhetoric of 'free' and 'fair' trade as well as 'making [childhood] poverty history', the rich world is still falling short in three key areas of international trade which are producing deadlocks in developing countries' economies: access to markets, which is hampered by pernicious trade barriers and tariffs for value-added agricultural products; continued subsidies; and the unjust dictates of commodity prices, in which poorer countries have no control.

The livelihood strategies of children within local economies are inextricably linked to national and global economic, social and political structures. The view that economic globalization is 'a tidal wave that will eventually raise all the boats' holds out the promise of improved living conditions for the poor. However, for children and families in Gedeo who are engaged in coffee production, the capitalist mode of economic relations seems to have produced only poison out of their coffee beans. Moreover, the implication of the 'tidal wave' metaphor, which, it is claimed, will change the lives of orphans, who must nonetheless compromise their well-being and development to meet their everyday needs, is equally telling. Again the paradox is that, in the wake of the AIDS epidemic, communities are grappling to cater for the basic needs of a huge number of orphans and destitute children. Sadly enough, they are doing this in the midst of extreme poverty in a wealthy world. The argument is that, alongside growing global wealth, we are seeing widening inequalities in children's living conditions, both within and between countries. What does globalization imply for the welfare of orphans and working children? How can global chains of care, support and solidarity be more effectively and efficiently coordinated so as to match those of global economic capitalism?

The experiences of orphans and working children in Gedeo suggest the need to extend debates on contemporary childhoods within the frameworks of post-development and post-structural theories. Children's livelihood strategies and living conditions in the North and South are quite disparate, and the structural circumstances under which they experience childhood are not the same. Children's agency is 'glorified' in contemporary, universal, advocacy-based discourses in which children are recognized as competent and independent social actors. However, the context in which they participate and exercise agency need to be elucidated and understood through empirical research. Although some of the children in Gedeo are capable of transforming their impoverishment into viable livelihood opportunities, their life chances are limited when compared with children who grow up in a context of overflowing choices and possibilities. Childhood in the former case is constrained by structural factors of poverty, inequality, debt, war, geo-political conflicts, epidemics, and ineffective legislation, which children in the latter cases are able to take for granted (Lund, this volume). Moreover, the crippling multiple effects of global trade on the livelihoods of families and children are shown in the current study. Normative views about the survival strategies of working children as the 'worst form of child labour that should be eradicated' is unhelpful from a policy perspective. In many societies, children have always worked and continue to work in order to ensure family livelihoods. Attempts to prevent children working simply reveal an understanding of children as the passive recipients of adult nurturing. Without compromising children's potential for social mobility, their work needs to be recognized as work in its own right, and as deserving appropriate material and social rewards at all levels, from local through national to global.

## Acknowledgements

I would like to thank all the children, families and other informants who participated in this research. I am grateful to the Research Council of Norway for funding the research and the fieldwork trips to Ethiopia on which the article is based. I also wish to convey my deepest thanks to the critical and constructive comments I received from the anonymous reviewers.

## Notes

1.  The fieldwork was conducted in 2005 (January–May) and in 2006 (January–March).
2.  The end of a managed market in 1989, together with US withdrawal from the International Coffee Association and major entrants like Vietnam into the market are the factors which have disrupted supply quotas and price levels in favor of wealthy coffee-roasting companies.
3.  Pseudonyms are used to maintain the anonymity of research participants.
4.  In 2001 alone, the total income that Ethiopia secured from the export of coffee slumped from US \$257 million to US \$149 million, as opposed to the \$58 million it had to save from debt relief.
5.  I borrow Collen O'manique's phrase 'globalization's pandemic' as a metaphor to denote the complex links and negative consequences of HIV/AIDS and unfair global trade on the lives of orphaned and working children.
6.  Annual Report of the Bureau of Education, Awassa: Ethiopia.

## References

Admassie, A. (2003) Child labor and schooling in the context of a subsistence rural economy: can they be compatible? *International Journal of Educational Development*, 23, 167–85.

Aitken, Stuart (2001) *Geographies of Young People: the Morally Contested Spaces of Identity*, London: Routledge.

Aitken, S. (2004) Global crises of childhood: rights, justice and the unchild-like child, *Area*, 33, 119–27.

Ansell, Nicola (2005) *Children, Youth and Development*, London: Routledge.

Ansell, N. and Robson, E. (2000) Young carers in Southern Africa: exploring stories from Zimbabwean secondary school students, in S.L. Holloway and G. Valentine (eds) *Children Geographies: Playing, Living, Learning*, London: Routledge, 174–93.

Ansell, N. and van Blerk, L. (2004) Children's migration as a household/family strategy: coping with AIDS in Malawi and Lesotho, *Journal of Southern African Studies*, 30, 673–90.

Ansell, N. and Young, L. (2004) Enabling households to support successful migration of AIDS orphans in Southern Africa, *AIDS Care*, 16, 3–10.

Aptekar, L. and Abebe, B. (2001) Conflict in the neighbourhood: street and working children in the public spaces, *Childhood*, 8, 477–88.

Bass, Loretta (2004) *Child labor in Sub-Saharan Africa*, Colorado: Lynne Rienner Publishers.

Bevan, P. and Pankhrust, A. (eds) (1996) *Ethiopian Villages Studies: Adado Gedeo*, accessed on: 12.03.05. http://www.csae.ox.ac.uk/evstudies/pdfs/adado/adado-hiphotos.pdf

Burman, E. (1995) 'What is it?' Masculinity and femininity in cultural representation of childhood, in S. Wilkinson and C. Kitzinger (eds) *Feminism and Discourse*, London: Sage, 49–66.

CSA (2000) *Demographic and Health Survey of Ethiopia 2000*, Addis Ababa: The Central Statistics Authority.

Ennew, J. (2002) Outside childhood: street children's rights, in B. Franklin (ed.) *The New Handbook of Children's Rights: Comparative Policy and Practice*, London: Routledge.

Ennew, J. (2005) Prisoners of childhood: orphans and economic dependency, in J. Qvortrup (ed.) *Studies of Modern Childhood: Society, Agency and Culture*, London: Palgrave, 128–46.

Escobar A. (1997) The making and unmaking of the Third World through development, in M. Rahnmea and V.Bawtree (eds) *The Post-Development Reader*, London: ZED, 85–93.

Grier, B. (1994) Invisible hands: the political economy of child labor in colonial Zimbabwe, 1890–1930, *Journal of Southern African Studies*, 20, 27–52.

Guest, E. (2003) *Children of AIDS: Africa's Orphan Crisis*, Pietermaritzburg: University of Natal Press.

Hamer, J.H. (1987) *Humane Development: Participation and Change among the Sidama of Ethiopia*, Birmingham, AL: University of Alabama Press.

Hecht, T. (2000) In search of Brazil's street children, in C. Panter-Brick and M.T. Smith (eds) *Abandoned Children*, Cambridge: Cambridge University Press, 146–60.

Hollos, M. (2002) The cultural construction of childhood: changing conceptions among the pare of northern Tanzania, *Childhood*, 9, 167–89.

Jones, S.P. (2004) When 'development' devastates: donor discourses, access to HIV/AIDS treatment in Africa and rethinking the landscape of development, *Third World Quarterly*, 25, 385–404.

Katz, Cindi (2004) *Growing up Global: Economic Restructuring and Children's Everyday Lives*, Minneapolis, MN: The University of Minnesota Press.

Lange, M. (2000) The demand for labor within the household: child labor in Togo, in S. Bernard (ed.) *The Exploited Child*, London and New York: ZED, 268–77.

Lund, R. (2007) At the interface of development studies and child research: rethinking the participating child, *Children's Geographies*, 5(1–2), 131–48.

Nieuwenhuys, Olga (1994) *Children's Life worlds: Gender, Welfare and Labor in the Developing World*, London: Routledge.

Nieuwenhuys, O. (2000). The household economy and the commercial exploitation of children's work: the case of Kerela, in B. Schlemmer (ed.) *The Exploited Child*, London: ZED, 278–91.

Nieuwenhuys, O. (2005) The wealth of children: reconsidering the child labor debate, in J. Qvortrup (ed.) *Studies in Modern Childhood: Society, Agency and Culture*, London: Macmillan, 167–83.

Nyambedha, E.O., Wandibba, S. and Aagaard-Hansen, J. (2003) Changing patterns of orphan care due to HIV epidemic in western Kenya, *Social Science and Medicine*, 65, 301–11.

OXFAM (2002) *Mugged: poverty in your coffee cup*, accessed on: 12.05.05 http://www.maketradefair.com/assets/english/mugged.pdf

Poluha, Eva (2004) *The Power of Continuity: Ethiopia through the Eyes of its Children*, Uppsala: Nordic Africa Institute.

Punch, S. (2001) Multiple methods and research relations with children in rural Bolivia, in M. Limb and C. Dwyer (eds) *Qualitative Methodologies for Geographers: Issues and Debates*, London: Arnold, 186–81.

Rita, C. (2001) Children's contribution to household labor in three sociocultural contexts: a southern Indian village, a Norwegian town and a Canadian city, *International Journal of Comparative Sociology*, 42, 353–67.

Robson, E. (2003) Children at work in rural northern Nigeria: patterns of age, space and gender, *Journal of Rural Studies*, 20, 193–210.

Robson, E. (2004) Hidden child workers: young carers in Zimbabwe, *Antipode*, 36, 227–48.

Robson *et al*. (2006) Young caregivers in the context of the HIV/AIDS pandemic in sub-Saharan Africa, *Population, Space and Place*, 12, 93–111.

Simmons, P. (1997) 'Women in development': a threat to liberation, in M. Rahnmea and V. Bawtree (eds) *The Post-Development Reader*, London: ZED, 244–55.

Stephens, S. (1995) Children and the politics of culture in 'Late Capitalism', in S. Stephens (ed.) *Children and the Politics of Culture*, Princeton, NJ: Princeton University Press, 3–44.

Tadesse Kippe (2002) *Five Hundred Years of Sustainability? A Case Study of Gedeo Land Use (Southern Ethiopia)*, Heelsum: Treemail Publishers.

Tatek Abebe (2002) Geographical problems and approaches to researching 'at risk' children, in G. Setten and S. Rudsar (eds) *Geographical Methods—Power and Morality in Geography Proceedings of the Annual Conference of the Norwegian Geographical Society*, Trondheim, 27–49.

Tatek Abebe (2005) Geographical dimensions of AIDS orphanhood in sub-Saharan Africa, *Norwegian Journal of Geography*, 59, 37–47.

UNDP (2005) *International Cooperation at the Crossroads: Aid, Trade and Security in an Unequal World*, http://hdr.undp.org/reports/global/2005/pdf/HDR05_complete.pdf, accessed on: 10.02.06.

Verhoef, H. (2005) A child has many mothers: views on child fostering from West Africa, *Childhood*, 12, 369–91.

Woodhead, M. (1999) Combating child labour: listen to what the children say, *Childhood*, 6, 27–49.

WHO (2000) *The World Health Report 2000—Health Systems: Improving performance*. Geneva: WHO.

# Negotiating Migrant Identities: Young People in Bolivia and Argentina

SAMANTHA PUNCH

*Department of Applied Social Science, University of Stirling, Stirling FK9 4LA, UK*

## Introduction

A key feature of globalisation is the notion of global interconnectedness (Kiely, 1998) where there is a blurring of boundaries between the local and the global. The global economy and global cultures are increasingly impacting upon children and young people's everyday lives (Kaufman and Rizzini, 2002; Katz, 2004). In particular, in the majority world, global economic restructuring has resulted in heightened levels of migration as people leave impoverished rural areas in search of employment and better lifestyles (Taracena, 2003; Schuerkens, 2005).

Young people in rural Bolivia have turned to migration as a way of coping with limited access to land and employment opportunities (Punch, 2002). Whilst this solves some of their problems, it also creates new ones. To some extent it widens their lifestyle choices for their future and it enables them to engage in a more global culture by providing them with the cash income required for a range of consumer goods. However, migration can also lead to increased pressures to obtain cash, perhaps instead of seeking further schooling (Bey, 2003; Aitken *et al.*, 2006) and can exacerbate inequalities between migrant and non-migrant households (Pribilsky, 2001; Carpena-Mendez, 2007).

In rural Bolivia, like many other rural areas of the majority world, education suffers from under-resourcing and poor teaching quality (Albornoz, 1993; Bey, 2003) as well as a lack of nearby available schools. Consequently young people have to migrate in order to continue with secondary schooling (Ansell, 2004). However, in rural Bolivia, when comparing the more tangible benefits received by migrating for work, migration for education seems less attractive for young people (Punch, 2004a). Thus the role of education as a site of social change and as a pathway to the increased status of an educated identity (Jeffrey and McDowell, 2004) is superseded by the role of migration in the lives of rural young people in Bolivia. This paper considers the role of migration and the movement of young people, goods and capital across the Bolivian–Argentinean border. It explores the ways in which a globalised capitalist system of agricultural production in Argentina offers young Bolivian migrants access to wider consumption practices compared with the traditional forms of subsistence production within their home community.

Furthermore, the paper discusses the ways in which migration provides young people with a source of collective and individual identity both within and outside of their rural community. On the one hand, young people face limited opportunities for work and are almost obliged to take on a migrant identity by leaving home to seek employment in the nearby town or neighbouring Argentina. On the other hand, many young people willingly seek the economic and social benefits of the migratory role. Thus, to some extent they are forced to join together through the economic constraints of their community, yet most of them perceive the migratory experience as an opportunity to enhance both their social and economic status, as well as facilitating their transition to adulthood. In particular, migration enables young people to consume more widely and access global goods as well as actively participate in constructing new opportunities for their future (see also Jeffrey and McDowell, 2004).

The paper begins by outlining the methods of the study followed by the social and cultural context in which young people live in rural Bolivia. It then highlights the nature of economic opportunities outside the community before exploring the ways in which young people perceive the opportunities and constraints of migration. Subsequently it discusses the freedom of the social world of young migrants during the four months or so when they return home. It argues that migration rather than education enables young people to develop their global interconnections and to increase their status back home.

## The Research Context

The paper draws on ethnographic research that I carried out in rural Bolivia, which explored how children and young people negotiate their autonomy in the main arenas of their everyday lives at home, at work, at school and at play (Punch, 2003, 2007). The study took place in the community of Churquiales, in the Camacho Valley of Tarija, the southernmost region of Bolivia. During the fieldwork, I lived for two extended periods in Churquiales (consisting of regular short visits over two years and a six-month intensive period of fieldwork when I lived with two households from the community[1]). The research also included two visits to north Argentina to conduct 20 semi-structured interviews with young migrants who had left the community in search of work.

The interviews with migrants included exploring their migrant history, their reasons for migrating, how they decided where to go, how they found work, the conditions of work, how they spent their earnings, the advantages and disadvantages of migrant work compared with staying within the community, and their future plans. It was important to conduct research with migrants both at home in their community and at the migrant destination as the immediacy of place shaped the ways in which they spoke about migration. For example, they tended to describe their migrant experiences through a rose-tinted lens when back in Bolivia but were more frank about the harsh living conditions when interviewed in Argentina.

In Churquiales there were a total of 58 households and I visited 18 of them regularly in order to conduct participant observation and interview all the household members. The sample of households was chosen to include households of different sizes with varied compositions such as young and old households, and those with mixed and single-sex siblings. Household visits lasted from half an hour to a whole day depending on the availability of household members (Punch, 2004b).

Churquiales is an economically poor and relatively isolated agricultural community with limited access to the mass media, as there is no electricity and no television, and communication networks are not extensive. The community is 55 km from Tarija, the regional capital, a journey of about four hours on the local twice-weekly bus. The main form of transport is on foot and there are no cars. The opportunities for waged employment are limited, and schooling is available only for the first six years of primary education (Punch, 2004a). Most of the families own two or three hectares of land, which they use mainly to cultivate potatoes, maize and a selection of fruit and vegetables. They also tend to own a small amount of pigs, goats and chickens, as well as a few cows. Most of their agricultural and livestock production is for family consumption, but any excesses are sold in local and regional markets. Consequently, as their household production is mainly traditional and subsistence-based, the labour requirements are high and all family members are expected to contribute to maintaining the household (Punch, 2001).

As already mentioned most households only own small plots of land which are not large enough to divide up between all their children. Since most families have approximately five children, and often as many as eight or nine, the family land is insufficient size for all the children to inherit. Thus young people are usually expected to seek alternative livelihoods at least for a while until some land becomes available. Consequently, in Churquiales, most young people do not own land, they have not yet inherited anything from their parents and they do not have their own personal capital to buy even the smallest of plots. The community offers few opportunities for permanent labour. Young people can work irregularly as day labourers for richer households but the daily wage is low and the work is very irregular. Their other main option is to migrate in search of work outside of the community, either to the nearby town of Tarija or, more importantly, to neighbouring Argentina which is approximately 100 km away (see Map 1). Thus migration also benefits those who stay behind by reducing stress on limited resources and land (Punch, 2007).

## Economic Opportunities Outside the Community

> At the moment I'm toying between whether to stay or go: if I go I like it, and if I stay I like it. Here I'm amongst friends and family, but in Argentina there are more economic opportunities.[2] (Lidio, 20 years)

When they reach about 15 or 16 years old young people consider going further away in search of higher wages and greater independence. Both girls and boys tend to search for migrant work after they have completed primary education and spent one or two years working for their household or within their community. By the time young people in Churquiales reach their twenties, most of them will have migrated at least once to Tarija and/or Argentina.

The region has a long tradition of migration to Argentina. Historically it dates back to the first half of this century when there was a need for a seasonal labour force on the

**Map 1.** Key migrant destinations include Oran, Pichanal, Fraile and San Salvador de Jujuy.

sugar plantations in North-west Argentina (Whiteford, 1975; Reboratti, 1996). Over the past several decades, an increase in citrus fruit and vegetable plantations, and the introduction of mechanisation on the sugar plantations, has meant that migration to Argentina has become more diversified, covering a wider area even as far south as La Plata region (over two full days' travel away). Migration to Argentina has also increased over the past three or four decades because of improved road communications and more developed social networks, facilitating employment contacts and opportunities (see also Schuerkens, 2005).

Thus, the capitalist nature of the large-scale agricultural production in Argentina has provided increasing opportunities for young people in Bolivia to migrate there seasonally each year. Migrant work in Argentina is particularly attractive to young, landless Bolivians because of the higher wages and the harvest period of the plantations coincides with the dry season in Bolivia when work opportunities at home are minimal. The agricultural labourer in Argentina earns an average of 10 dollars daily whereas the daily wage in the communities of the Camacho valley in Bolivia is 10 *bolivianos*, which is equivalent to two and a half dollars, at least four times lower than that in Argentina. However, it is worth noting that despite this being a good wage in comparison with Bolivia, it is lower than an Argentinean would expect for the same work. It is thus in the Argentinean landowners' interest to employ Bolivian migrants and this increases the transnational interdependencies between the employers and employees. For young Bolivians there can be little incentive to work in their community for low agricultural wages when Argentina is less than a day's bus journey away. Hence those more likely to migrate on a regular basis are those with less land or landless young people in search of a cash income to satisfy their growing financial needs in the transition to adulthood (Punch, 2002; Carpena-Mendez, 2007).

Some girls go to Tarija to work as domestic maids and some boys work in construction or some form of manual labouring, such as a carpenters' apprentice, in Tarija. However, most boys are likely to work on commercial farms in the north of Argentina and some girls travel with male companions to undertake the same agricultural work. Other girls may accompany male relatives to cook and carry out domestic duties whilst the men work in the fields. Girls tend to migrate to Argentina later than boys, often after an initial period of working in Tarija.

Girls who live in as domestic workers in Tarija are paid about 100 bolivianos (25US\$) until they have learnt to cook and clean in an urban household, and then their wage rises to 150–200 bolivianos (38–50US\$) a month plus food and lodging. For this they work a six and a half-day week with only Sunday afternoons off. They rarely go to Argentina until they are 19 or 20 years old, unless accompanying a close relative. This is largely because of parents' beliefs that girls have to be looked after whereas boys do not have to be so protected (see also Chant and Radcliffe, 1992). The proximity of Tarija means that parents can visit regularly and the young people can also return to Churquiales frequently. Girls may have to face greater negotiations with their parents when seeking migrant work. As Vicki's mother said to her when she wanted to go to Argentina: 'No, we worry too much. We want you to work in Tarija where you are nearer home'.[3]

Boys of 15 or 16 years tend not to pass through the intermediate stage of working in Tarija, but go further, crossing the border to Argentina. Their first trip is usually with a friend or relative who has been before, thus ensuring a better chance of securing employment. They also earn a lower wage at first, \$7–8 a day instead of \$10–12, whilst they learn the regional agricultural practices and how to use the different tools. The rest of this paper focuses mainly on the experiences of young people who migrate seasonally to Argentina: boys aged 15–25 and girls aged 19–25.

These young people migrate seasonally each year, leaving their community between March and May and returning after the harvests in Argentina in November and December. Consequently they spend approximately six to 10 months away working on the commercial farms in Argentina and two to six months back at home in their community. During their time away both young male and female migrants work long hours, have limited free time and their life revolves around their work. When they come back to their community they consider themselves on holiday, to have a rest, though they do help out at home during this time. Thus their lives are basically divided into an economic world of work in Argentina and a social world of leisure in Churquiales.

## Migrant Work: Constraints and Opportunities

The main reason why young people migrate is the lack of economic opportunities in Churquiales:

> I came to Argentina to work, in Churquiales you can't earn anything, I also came to learn how to work on the plantations. The money is for me and for my family, to help my dad. There's no work in Churquiales, I've worked there lots for my family, but they don't pay me. Though in the future I want to go back there.[4] (Domingo, 14 years, Pichanal, Argentina)

This coincides with other migration studies in Bolivia which have found that the most important reasons for out-migration are the lack of access to land and limited sources of paid employment (Pérez-Crespo, 1991a) as well as young people's desire for consumer goods (Pérez-Crespo, 1991b). However, there are disadvantages as the migrants are expected to work extremely hard, long hours in the heat amongst flies and mosquitoes. Migrant work is tiring with little free time and a maximum of one day off a week. Living conditions are basic for migrants, often requiring them to sleep on the floor in cramped conditions, usually in a hut near the fields where they are working (see also Reboratti, 1976). The cost of living is high, so to survive without spending all their earnings they have to eat basic foodstuffs:

> I didn't spend much money in Argentina and I don't spend much in Bolivia either because it's better to save money. It's not worth while going to Argentina for a short time due to the transport expenses, you've got to go for at least two months minimum. Just as you can earn money, you spend it on food and transport.[5] (Lidio, 20 years)

Hence, young migrants prefer to live basically, with little or no comfort, in order to accumulate savings for their return to Bolivia. On the large scale agricultural plantations the conditions of work are far from ideal yet the migrants still welcome the economic opportunities they offer (see also Sklair, 1994).

The work may be arduous but the financial gains can be substantial. One young migrant explained how he took advantage of being able to work long hours in order to earn a high income. He said it was extremely tiring, but he felt satisfied that the hard-work was rewarded with good pay:

> I work as a pieceworker travelling around the different areas. In February it's the grape harvest in Mendoza. In the north between April and September there are the tomato and pepper harvests. From August onwards in Corrientes there are the tomato and citrus harvests. Between December and 15 January it's the tomato harvest in Buenos Aires. After September the harvests in the north come to an end because the heat doesn't let you continue working. I go to different places depending on the harvest season, it's a great way of seeing places and meeting people.[6] (Julián, 21 years)

He preferred the flexibility of doing piecework as it allowed him the freedom to work as much or as little as he liked and to move between the different harvests. Migrant work gives young people access to a higher income and economic opportunities which are lacking in their home community:

> The first time I came to Argentina was to look for work, because apart from that, Tarija is a great place to live, it's cheaper, it's nicer, it's better all round. My brothers came about ten years before I did. I came with them the first time and then I kept coming. Nearly all the young blokes come here. In Tarija there are no jobs unless you've been to university. We could cultivate a hectare of tomatoes in Tarija, because it produces really well, but you can't sell a kilo for tuppence, it's not worth anything, that's the problem.[7] (Dimar, parent, Fraile, Argentina)

The influence of return migrants is particularly strong for young people. Gradually the style of dress of the migrants changes. For example, instead of open sandals and trousers, they wear shoes or trainers and jeans (see Figures 1 and 2). During local celebrations children see returned migrants well-dressed with new fashions and they say: 'I want to go too, I'm going next year'.[8] The return migrants displaying their new clothes and material goods

**Figure 1.** Young migrants with their Argentinean shirts, jeans and denim jackets.

(such as steroes) provide other young people with information about global youth cultures and act to some extent as an 'opening into the modern world' (Bey, 2003, p. 295). This is particularly relevant in a relatively isolated rural context where access to the media and images of life outside the community are extremely limited. Thus, since most young people return from Argentina or Tarija with something to show for their hard work, others are also inspired to leave.

> Since all the blokes of that age came to Argentina, you just wanted to go too, I was 16, just a kid. All my friends came here and then went back saying 'Oh, it's just great there', so I just wanted to come here too.[9] (Juvenal, parent, Fraile, Argentina)

> The young person leaves to earn his own money to buy clothes and to go drinking, here there aren't any opportunities.[10] (Pedro, parent)

Many migrate initially out of curiosity and because they do not want to be left behind. Exaggerated stories of how great a life the migrants lead encourages others. Migrant employment supplies young people with money for their growing personal needs and social activities, which most parents are unable to provide. At the same time, remittances sent home or money brought back at the end of the season contribute to the family income. This can be a very important source of cash for households which are largely reliant on a subsistence-based rural economy (see also Lund, this volume).

Some young migrants expressed family disputes as a motive for leaving for the first time, as a chance to escape from family problems temporarily. For example, Hugo was 16 years old when he went to Argentina with friends without telling his father. He left

**Figure 2.** Young migrants at a community party.

suddenly as a way of escaping from his father who had been getting drunk and beating him over the previous few months.

Migrating while young, with limited knowledge of life outside the community, can be an exciting or a bewildering experience: 'I always thought it was going to be great, but at first it was difficult to get used to it'.[11] Leaving their sheltered, relatively isolated community for the first time can be difficult. Migration requires processes of adaptation (Schuerkens, 2005) as the young migrants gradually adjust to the new environment, learning ways to overcome the difficulties:

> There I learnt the cost and value of life. Previously my parents gave me everything: food and clothes, etc. And there I had to work for these things.[12] (Gerardo, 20 years)

Seeking work in Argentina is facilitated by an established migratory tradition, which has been built up over several decades. This not only encourages new migrants to leave but also facilitates their initial move to live away from home:

> As I'm going with friends and I'll know many who are there, I don't think it's going to be very difficult. Going alone would be hard and I'd miss here a lot, but with so many people I know, I don't think so.[13] (Sebastián, 18 years)

Ansell and van Blerk (2007) remind us that the emotional impacts of migration are often under-explored. They argue that the process of 'getting used to' a new place, in particular building familiarity and a sense of belonging, is facilitated by relationships with family and friends in the migrant destination. Social networks are extremely important for the young Bolivians migrating for the first time. The majority travel with experienced migrants or have relatives or friends in the area of destination who can assist them in finding work and accommodation. Most expect a maximum of two weeks before they find work. In many cases, their first job is arranged before they even leave Churquiales, through a relative or friend who has already established links. The contact links between migrants is crucial in providing an extensive migrant network which facilitates the entry of newcomers (see also Roberts, 1995; Schuerkens, 2005). This makes a potentially overwhelming experience more enjoyable and less stressful.

Entry into Argentina can be problematic for migrants. They develop strategies for crossing the border with no documents since the official documentation is expensive and difficult to obtain. For example, they cross through the forest and not at the controlled point of entry, or they obtain a tourist pass for a few days and do not return within the stipulated time. Sometimes they borrow documents from relatives. Alternatively, as one migrant said: 'I go at night when they don't check much. I go with just one bag'.[14] He takes minimal luggage so it is not obvious that he is going for a long time. Once in Argentina the migrants can work for a few months to save sufficient money to be able to acquire a passport and working visa. If they are refused entry, rather than return home, they usually find work on the sugar-cane plantations in Bermejo on the Bolivian border.

Many young rural people are contracted for work in Argentina whilst they are still in Bolivia:

> There are some fellow countrymen who go to Tarija and contract twenty or thirty workers, as labour is so cheap, so cheap that they take advantage of it. They make the contracts in Bolivia . . . Then they bring the workers to Argentina and they have to work their guts out, while being paid a Bolivian wage. I reckon that's bad, but lots do it. The contract workers don't have any freedom, they work from morning

to night, and earn a Bolivian, not an Argentinean, wage.[15] (Vicente, parent, Santa Rosa, Argentina)

Whilst such an arrangement can be considered exploitative, it also has advantages:

The contract was made in Bolivia, related to the Bolivian wage level. I worked in Argentina but was paid in Bolivia. It suited us to be able to come here, I liked it and returned the next year alone.[16] (Juvenal, parent, Fraile, Argentina)

A set contract can be beneficial to a new migrant as it guarantees employment, covers transport and food costs, enables them to cross the border and offers security. Since contract workers are paid on return to Bolivia, their wage is intact for their own use during the summer holiday and for their family. A 14-year-old migrant from Churquiales returned with US$600 after his four-month contract period. He bought a stereo cassette player for himself, a wardrobe for his parents and merchandise for the household. If such migrants decide to return to Argentina, most do so alone. The initial experience enables them to make contacts to find their own independent job for the next time, allowing them more freedom and a better wage.

Young people's perceptions of migrant work in Argentina vary: some love it, some hate it and some are ambivalent about it:

Some don't like it there because of the heat or the flies, or because of the work: tough, hard and tiring. But I like it all. I've never said that I don't like it.[17] (Domingo, 14 years)

Domingo was always determined to go, but his older brother, Lorenzo, was more undecided about the experience. Their mother explained:

Domingo is harder, he doesn't feel sad, just a bit at first and then he got used to it quickly. He's very independent now but he's always been like that. Whereas Lorenzo gets more upset.[18] (Dolores, parent)

It was also common for young people to change their minds about migration. For example, Alcira's brothers asked her to go with them to Argentina as their cook, but she did not want to because: 'I wasn't used to leaving my house'.[19] She subsequently worked with her sister for several months in Tarija and now hoped to join her brothers to Argentina on their next trip. This again reinforces the importance of social networks of family and friends in easing the emotional impacts of migration (Ansell and van Blerk, 2007).

Therefore, young people's decision whether to migrate is based on a range of opportunities and constraints: push and pull factors at both the sender community and the destination (Lee, 1966). At home the main push factors are a lack of access to land and limited wage opportunities, and the pull factors, which may restrain the young person, include family obligations. The main pull factors at the destination are higher wages and the opportunity to travel, whereas the main push factor discouraging the migrant is unfamiliarity with the destination (Punch, 2007). As Lee (1966) indicates, ultimately the incentive to leave has to be higher than that to stay for young people to decide to migrate.

## Young Migrant's Social World in Churquiales

Whilst the migrant work can be tough, mostly it is seasonal and young migrants usually get to spend about three or four months back home in their community after the harvests in Argentina and before the planting begins for the following season. They have limited time and opportunities for indulging in an active social life in Argentina or Tarija, so they make the most of the time they are in Churquiales by going to many social events and parties. One return migrant explained:

> There's not much social life over there ... It's more work and then we come here to enjoy ourselves. There it is expensive and each day we have to turn up for work, but here we are free from all that, we just have to help our parents a bit.[20] (Martín, 19 years)

In Churquiales *fiestas* were usually arranged for a specific occasion such as: a particular Saint's Day, a birthday, a *hierra* (cattle marking day), All Saint's Day, Christmas, New Year, Carnival, Easter, *Chaguaya* August festival and farewell parties. As well as going to parties, young people's social life included going to football matches. Games were organised against nearby communities and it was expected that the losers bought drinks for the winners. Matches were played on Sundays and were followed by a drinking session in the evening. Young women also attended the football matches, as this was when all young people gathered together to chat and socialise regardless of whether they were actually playing in the match or not. Girls did not merely go to watch the boys play from the sidelines, they used it as an opportunity to get together with their friends, in much the same way that not all the boys who went were playing in the match. They went because it was a social occasion, but one which revolved around a male-centred activity. This coincides with the research of Matthews *et al.* (2000) who found that public outdoor spaces can provide an important social venue for girls as well as boys.

Young people who were beginning to go to *fiestas* had an increased demand for more fashionable and smart clothes to wear compared to when they were children. Young people in rural Bolivia experience a similar desire for consumerism like many young people in the minority world (Jones and Wallace, 1992; Liechty, 1995). Increasing social activities meant that not only did they want to look good with their peers and dress appropriately, but they also needed money to spend at such events. On a Sunday young people wanted money to buy something to eat or drink during the afternoon. Young men particularly tended to want to have money to be able to bet on the winning team, or to gamble in a card game. At *fiestas* they wanted to be able to pay for a jug of wine to share with their friends.

Young people needed access to economic resources to be able to take part in such social activities and have access to a wider consumer market to buy the requisite clothes and shoes, as indicated in the following quotations:

> Argentinean clothes are better quality. We dress like they do. ... Clothes here are nothing like clothes from there. Any old top there costs 5–6 dollars, but one of a particular brand costs 25 to 30 dollars.[21] (Maria, 23 years)

> I buy clothes there because they are beautiful, and here because they are cheap.[22] (Luisa, 19 years)

In the countryside, where financial resources were limited, parents would rarely pay for young people's fashionable clothes or social activities. Therefore, acquisition of greater social autonomy tended to be linked to economic independence for young people, and their entry into a more independent social world largely depended on their simultaneous entry into paid employment. As young migrants, they had left home (albeit temporarily) to work and were now more economically independent. In addition, they considered themselves more socially independent as they had lived by themselves while working, so they did not appreciate being restricted by parental rules when they were home after a period of freedom from parental authority:

> Many young people leave home to go and work somewhere else, and when they come back they don't take any notice of their parents. They do what they want. They think that it's enough to give their parents money and that now they are independent. Look at Tomás' daughters—buying and drinking wine with boys sitting opposite them. And Roberto went off for days drinking . . . Some still do what their parents say, but it's rare. Like Pedro's children: they are like ants in his house. They're scared of him and they do what he wants.[23] (Gertrudes, parent)

Not only were they more economically independent, but they also contributed to their household, usually in the form of money sent back while they were away, and money and goods brought back on return to the community (see also Lund, this volume). The importance of the young migrants' economic contribution to their household meant that they tended to acquire greater social rights, such as the right to decide when they would go out, with whom, where and until when. They were bringing money into their household, so parents were less able to dictate what they did, and the young migrants asserted their autonomy: making their own decisions, and controlling their own time and use of space. However, as they were still living with their parents during this time back in the community, to some extent they were expected to respect some of their parents' wishes, such as helping out with certain jobs or letting them know when they would be in for a meal. As the above quotation illustrated, parental control varies between households. Some parents are stricter and still maintain restrictions over their children's behaviour. Nevertheless, most parents realise that by this stage their control over their children has decreased. If they do not like their child's behaviour and force them to conform, then the young migrant may be tempted to return sooner to Argentina.

Most adults in the community, whether their children migrate or not, criticise the young migrants' actions during their stay over the summer months. In particular, there are tensions between non-migrant and migrant households. Non-migrants resent the migrants showing off and treating those who do not migrate as if they were inferior:

> They don't want to speak with their own people any more because they think more of themselves. They come back refined. . . . The work is just the same as here: agriculture. And the girls are in houses scrubbing floors. But because they earn more and they buy themselves clothes, they come back here thinking they're better. Some aren't like that but many young people are. The older ones are more used to it.[24] (Manuela, parent)

> The only thing they learn is to speak with a funny accent. They come back and show off their riches. And they forget to say that they worked like slaves there and didn't eat very well.[25] (Felicia, parent)

The young migrants have earned themselves a reputation for being mainly interested in going to parties, getting drunk and having a good time in Churquiales. Many adults criticised such social freedom which they considered as too exaggerated and indulgent (Punch, 2007). They disapproved of the migrants wasting all their earnings on drink and festivities:

> Most of them spend their money just on clothes and drink. They don't think about the future, and they will always have to return there (to Argentina) to work.[26] (Tomás, parent)

The non-migrants scorned how most of the migrants' earnings seemed to be spent purely on temporary social activities rather than invested in buying cattle or saving up to buy some land or build a house. However, for the young people this period of their lives represents a time of immense freedom and autonomy:

> We like coming here for the summer: to enjoy ourselves and ride around on our horses. To spend one month, up to three months here, it's nice in summer. In summer it's all green and there are parties.[27] (Gerardo, 20 years)

They experience control over their own lives and have access to consumer markets and social power. This coincides with Sklair's observation that:

> . . . poor people, in poor and rich countries, apparently defy economic rationality by purchasing relatively expensive global brands in order to forge some sense of identity with what we can only call in a rather crude sense 'symbols of modernity'. (Sklair, 1994, p. 179)

To some extent the young migrants' attitudes are not surprising in a climate of neoliberal globalisation where some Latin American governments promote 'the idea that consumerism is synonymous with liberty' (Lacy, 2002: 124).

Young female migrants also take part in consumer culture. However, since the majority of them tend to migrate to Tarija, their earnings are not as high as those who work in Argentina. Young women also tend to drink less than young men, spending more of their earnings on clothes to wear at parties, rather than alcohol. Vicki, who worked as a maid in Tarija, explained that she bought clothes, a cassette player, camera and second-hand bicycle with her migrant earnings. She also saved money to spend at parties in Churquiales during the holidays:

> I would rather not go to a party if I don't have anything to wear. I can't go to two parties in a row wearing the same dress, people would look down on me. That's why from August onwards I save hard until Christmas. It's also because my friends come back from Argentina, waving their money around. So when everyone chips in at a party, I can also give something rather than face the embarrassment of having to say I don't have any money.[28] (Vicki, 17 years)

As children they have had limited consumer power to have the goods and clothes they desire, so when young people have the opportunity to buy what they want, they indulge in conspicuous consumption. Besides, since they have to work extremely hard in Argentina or in Tarija, they feel they deserve to spend their earnings as they wish: 'I like it here. It's great for the summer, for drinking and going to parties. It's good not having to work. Nobody bothering you'.[29]

By helping their families, by buying goods and gifts, they gain the freedom to decide how to spend the rest of their money, but adults can be quick to criticise them. One young migrant explained that 'People here are envious'.[30] He felt that many non-migrants envy the migrants' economic success and social freedom. Since many young people migrate from quite a young age (most go first between 15 and 19 years) then they can afford to 'waste' at least a few of the initial years by indulging themselves rather than worrying about the future. However, such conspicuous consumption can lead to inter-generational tensions as adults express their concerns of the young migrants' behaviour. Their greatest criticism is that they drink until all their money has run out: 'They come back for some good holidays and when they don't have any money left, they go off again'.[31] However, the young migrants have chosen to live for now in the present rather than invest in the future. As they get older, they tend to become more responsible and gradually begin to save to buy animals, land and a house when they start to think about having a family of their own. However, not all migrants will be able to accumulate enough savings to buy their own land: some will inherit land from their parents or may sharecrop with other households, and others will continue to migrate either seasonally or permanently.

It was during the return visits home that the migrants would meet and learn from one another's migratory experiences. Many of the male migrants would share their knowledge about the economic opportunities on different commercial farms. Similarly, the female migrants made contacts for securing domestic work in the towns. Sometimes they would team up with the male migrants: either to work on the farms or to join them as their cook. The importance of the young migrants' social life at home meant that they tried to come back at times coinciding with local festivals. Most young migrants aimed to be back for the celebrations of All Saints' Day on 1–2 November, and stay throughout Christmas, New Year and at least until Carnival (in February/March, just before Lent). If possible they also tried to come back for a break in August to make the most of the festival of *Chaguaya*. The migrants' return was therefore largely influenced by *fiestas*: 'Now he wants to go back to Argentina because the *Chaguaya* festival is over and he's also spent all his money'.[32] The migrants have a great impact on the community's social world, since many of them return more or less at the same time each year, the community comes to life, full of young people and festivities, during those summer months. The migrants wish to make the most of their 'holiday', so they actively organise social events such as *fiestas* and inter-communal football matches. Thus the social world not only shapes their actions, but they also play an active part in shaping their social world.

## Young Migrant Identities

Unless young people's parents lend them land, or until they have inherited or bought land, their work opportunities in Churquiales are limited. The initial decision whether to stay or leave is complex. On the one hand, there is an inquisitive desire to travel, discover new things, 'see what it's like' and not be left behind (Bey, 2003). On the other hand, there is the apprehension of leaving home for the first time, leaving the comfort and familiarity of being in one's own environment with family and friends (Ansell and van Blerk, 2007). Each young person has to consider the benefits and drawbacks of staying or leaving according to their individual situation and personal preferences. For example, Bertha had worked as a domestic maid in Churquiales and Tarija. She decided she preferred to work in Tarija: 'it's less work, just in the house and it pays more'[33] Similarly, Laura preferred working in urban areas: 'here the work is lighter, only in the kitchen. There's no animals to look after'.[34]

Young Bolivians also attempt to strike a balance between striving for more independence, whilst considering household needs and family responsibility. Elsewhere I have referred to this as 'negotiated interdependencies' between and within the generations involved in family relationships (Punch, 2002; see also Kjørholt, this volume). Furthermore, to some extent the migrant decision depends on young people's position in the birth order as older siblings tend to be able to migrate at a younger age (Punch, 2001). In contrast, the youngest siblings tend to experience more pressure to stay and care for their parents (see also Nieuwenhuys, this volume). Ultimately young people's decision is made within a range of opportunities and constraints, which involve a consideration of dependencies and independencies, family and work contexts, and communities and individuation (see also Aitken, this volume).

Chawla (2002) argues that a combination of increased consumption in the minority world, growing populations and rising expectations in the majority world have led to increasing competition for natural resources, in particular land. She notes that rural children are more likely to be involved in commercial agricultural production rather than subsistence farming. Churquiales is a community in gradual process of 'modernisation' as the agricultural production continues to be mainly subsistence-based with small surpluses sold in local and regional markets. However, migrant work in Argentina is more commercialised and thus migration leads to new forms of agricultural learning for young people. Whilst this can encourage young people to develop these skills back in Bolivia, it can also lead to tensions between old and new knowledge (see Katz, 2004).

In this rural Bolivian community, migration is an important part of youth identity. As Aitken (this volume) argues, young people's work and identity is empowering because of the contexts in which they are embedded rather than the work itself. Despite the poor conditions of work in Argentina, the young Bolivian migrants have substantial social freedom when back home in their community as well as greater access to consumer products. Their migrant employment offers the opportunity to enhance their individual social status and also to contribute to their households and maintain interdependent family ties. Furthermore since migrants return with their newly acquired 'symbols of modernity' (Sklair, 1994), they impact upon the aspirations of younger children in the community.

In contrast, there are only a minority of young people in Churquiales who have continued with their formal education and in many cases secondary education is not perceived as a viable alternative to the more immediate material gains of work and migration (Punch, 2002). As Bey argues: 'along with the economic function of work there is an element of learning and apprenticeship, which provide a preparation for entry into the adult world' (2003, p. 295). In keeping with other majority world researchers (Bey, 2003; Katz, 2004; Aitken *et al.*, 2006), the informal local education gained through work either within or outside of the community provides young Bolivians with practical and appropriate skills which are more relevant for their future livelihoods than those gained through formal school-based universal education (Punch, 2004a).

Migration is a way for young people to cope in this context of limited agricultural land and a lack of employment opportunities, and is therefore a relatively rational pathway for them to pursue. An outcome of this is that young migrant identities represent financial gain and enhanced social autonomy, thereby encouraging more young people to migrate. Migration increases their economic and social capital and allows them to be more flexible in their choice of youth transition (see Bynner, 2001). They may decide to migrate on a seasonal basis, to return to their community or to live more permanently in Argentina or Tarija. Migration is often used by young people as a bridge between being part of their parents' household and forming a new household of their own (Punch, 2004a). It allows them to start to accumulate savings whilst also maintaining links with their parents and siblings by sending remittances home. Bosco (this volume) refers to this as 'networked geographies of responsibility' as people care for others at a distance.

On the one hand, migration is a practical requirement because of the lack of economic opportunities within their own rural community, yet on the other hand, it enables young people to 'strategize in relation to their futures' (Jeffrey and McDowell, 2004, p. 134). In this sense, regarding whether they migrate or not, young people's agency is bounded (see also Ansell, 2004). However, their migrant experiences are shared during their return visits home and provide them with a sense of collective as well as individual identity. They arrive back at their community with new material goods as symbols of their increased prestige and enhanced ability to consume in a wider range of markets. It is during these summer months back home when they not only inspire other young people to migrate but also have a substantial impact on the social life of the community. Thus, migration is a coping strategy which facilitates young people's participation in a more consumer, and sometimes global, culture as well as enhancing their social and economic autonomy. It offers young people access to multiple and complex migrant identities which they negotiate at an individual, household and community level. However, migration can also lead to growing economic inequalities within the community and increased tensions between migrants and non-migrants (see also Pribilsky, 2001; Carpena-Mendez, 2007). As Taracena (2003) found in Mexico, the migrant identity can open up new horizons but can also result in difficulties when losing or adapting traditional lifestyles and incorporating more 'modern' behaviours.

It may be argued that for some young people in rural Bolivia their migrant status becomes a key feature of their youth identity as it influences not only how they see themselves but also how they are viewed by others in the community and at the migrant destination. However, it is worth remembering that migrants are not a homogenous group and they will blend their 'experiences of multiple places simultaneously' (Silvey and Lawson, 1999, p. 125) but also in diverse ways. As Beazley (2003, p. 18) argues in relation to street children in Indonesia, they have collective and multiple self-identities and they '... have to negotiate their identities and adapt their activities and strategies in response to their changing environments'. Similarly, many Bolivian migrants express ambivalent feelings about the process of migration, and their perceptions may change over time: 'Thus migrants' identities are reworked through the nature of the destination in which they find themselves' (Silvey and Lawson, 1999, p. 125). Migration is foremost a survival strategy but a byproduct of this is the development of collective and individual migrant identities which young people negotiate both within and outside the community.

Furthermore, it is worth bearing in mind that the migrant identity is a precarious one as it is largely dependent on the global economy. Neoliberal globalisation has impacted negatively on both the Bolivian and Argentinean economies (Lacy, 2002), and Argentina suffered a financial crash in 2002 which led to extreme economic hardship and social chaos (López Levy, 2004; see also Bosco this volume). Further research is now needed to explore the knock-on effect of this economic crisis for the thousands of Bolivians who migrated there seasonally each year. This is a reminder that processes of globalisation can impact upon children and young people in unpredictable and unforeseen ways (see also Kaufman *et al.*, 2002). There are also implications for processes of development (Schuerkens, 2005), and perhaps Bolivia needs to develop a more viable agricultural economy (see Chawla, 2002) or promote local farming and the ruralisation of primary education (see Meinert, 2003) in order to encourage more young people to stay in rural communities rather than migrate.

## Notes

1. This doctoral research developed as a result of initially working for two years in the same region of southern Bolivia on an European Union funded project (managed by Dr David Preston) titled: *Farmer Strategies and Production Systems in Fragile Environments in Mountainous Areas of Latin America.* The names of the community and the respondents have all been changed in order to maintain anonymity and confidentiality.
2. Ahorita estoy balanceando: si voy allá me gusta, y si estoy aquí me gusta. Aquí estoy dentro de conocidos y estoy con la familia. Pero allá hay más oportunidades económicas.

3. No, nos preocupa demasiado. Queremos que trabajes en Tarija donde estás más cerca de la casa.

4. Yo he venido a trabajar, allá no se gana nada, también he venido para aprender a atender almácigos. La plata es para mí y para la casa, para ayudar a mi papá. No hay trabajo en Churquiales. He trabajado mucho allá para la casa, pero no me pagan. Pero en el futuro quiero volver a Churquiales.

5. Allí no he gastado mucho y aquí tampoco porque es mejor ahorrar dinero. No vale la pena ir a Argentina por muy poco tiempo por los gastos del viaje. Hay que ir durante dos meses mínimo. Como se gana se pierde también en la comida y el transporte.

6. Trabajo como tantero recorriendo las zonas diferentes. En febrero es la cosecha de uva en Mendoza, en el norte de abril hasta septiembre hay cosechas de tomate y pimiento, a partir de agosto en Corrientes hay cosecha de tomate y citrico, entre diciembre y el 15 de enero hay cosecha de tomate en Buenos Aires. A partir de septiembre las cosechas en el norte terminan porque ya el calor no deja trabajar. Recorro los lugares depende de las épocas, es forma de conocer lugares y gente.

7. La primera vez he venido por motivo de trabajo, después Tarija es lindo para vivir, más barato, más todo, más lindo. Mis hermanos venían antes que yo, unos diez años antes, y yo he venido con ellos la primera vez, y seguía viniendo. Toda la juventud viene casi, y en el mes de diciembre vuelvan todos a su comunidad y lindo está, en camionetas van. En Tarija no hay en que trabajar, si uno no es profesional. En Tarija nosotros podríamos poner una hectarea de tomate, y se podría hacer rendir lindo, pero no se puede vender un kilo ni en un centavo, no vale nada, eso es el problema.

8. Yo quiero ir también, voy al año.

9. Como todos los muchachos ya de esa edad venían, uno ya quería venirse, yo tenía 16 años, era chico. Todos los compañeros venían para acá y decían: 'ah allá es lindo', entonces uno ya quería venirse.

10. El jovencito sale para hacer su plata para su ropa y para tomar. Aquí no hay oportunidades.

11. Siempre pensaba que iba a ser lindo pero al principio fue difícil acostumbrarse.

12. Ahí he aprendido el costo y el valor de la vida. Antes mis papás me daban todo: comida y ropa, etc. Y ahí tuve que trabajar para estas cosas.

13. Como estoy yendo entre amigos y boy a conocer muchos que están allá, no creo que va a ser muy difícil. Yendo solito sería duro y extrañaría harto pero con tantos conocidos, no creo.

14. Yo voy por la noche cuando no revisan mucho. Sólo con un bolsón.

15. Hay unos paisanos que se van a Tarija y contratan 20 o 30, como ahí la mano de obra es tan barata no? Tan barata que se aprovechan. Contratan allá, por esa plata de allá y les traen y les revientan laborando acá. Pero pagando con sueldo boliviano. Eso yo lo veo mal, y muchos hacen eso. No tienen libertad de trabajo, ellos vienen y les hacen trabajar desde la mañana hasta la noche, por la misma plata, pero la de allá, no de acá.

16. El contrato ha sido allá en Bolivia, el tanto y el precio, todo. Era relacionado con la moneda boliviana, trabajaba aquí pero me pagaban allá. Nos convenía para venirnos. Me gustó, fui de nuevo a Bolivia y de allá ya vine sólo.

17. A algunos no les gusta allá por mucho calor o mucha mosca, o por el trabajo: fierro, duro y pesado. Pero a mí me gusta todo. Yo nunca he dicho que no me gusta.

18. Facundo es más seco, no le da pena, sólo al principio un poco y después se acostumbró rápido. Es muy independient ya pero él siempre ha sido así. En cambio Jacinto es más penoso.

19. No sabía salir de mi casa.

20. No hay mucha vida social allá ... Pero más es el trabajo y aquí venimos para divertirnos. Allá es caro y diario tenemos que presentarmos al trabajo, pero aquí somos libres de todo esto, sólo ayudamos un poco a los papás.

21. La ropa argentina es de mejor calidad. Nos vestimos a la manera que se visten allá. ... La ropa aquí no tiene nada que var con la ropa de allá. Una polera cualquiera allá cuesta 5–6 pesos, pero una de marca cuesta 25–30 pesos.

22. Compro ropa allá porque es muy linda y aquí porque es más barata.

23. Muchos jovenes se salen de sus casas a ir a otro lado a trabajar, y vuelven y no hacen caso a sus papás. Hacen lo que quieren. Piensan que basta dar plata a sus papás y ya se independizan. Mira las chicas de Tomás—comprando y tomando vino con los chicos enfrente. Y Roberto se quedaba días por ahí a tomar ... Algunos todavía hacen caso de los papás, aunque rara vez. Como los hijos de Pedro: son como hormigas en su casa. Le tienen miedo, y hacen lo que él quiera.

24. Ya no quieren hablar con su gente porque se creen más. Vuelven refinados ... El trabajo es el mismo que aquí no más, en agricultura. Y las chicas en casas fregando pisos. Pero por lo que ganan más y se compran ropa, ya vuelven aquí pensando que son mejores. Algunos no son así pero muchos jóvenes sí. Los viejos ya se acostumbran.

25. Lo único que aprenden es hablar con un accento raro. Vuelven y demuestran sus riquezas. Y olvidan a decir que trabajan allá como esclavos y no comían bien.

26. La mayoría gastan en ropa y trajo no más. No piensan en el futuro y siempre van a tener que volver allá a trabajar.
27. Nos gusta venir para pasar verano: divertirnos y pasear en caballo. Para pasar un mes, hasta tres meses aquí, es lindo en verano. Verano es verde y hay las fiestas.
28. Prefería no ir a una fiesta si no tengo nada para poner. No puedo ir a dos fiestas seguidas con el mismo vestido, porque la gente mira mal. Por eso desde agosto ahorro fuerte hasta navidad. También es porque mis amigos vuelven de Argentina, mostrando su plata. Entonces cuando todos dan una vaca en una fiesta, yo también puedo dar, en vez de tener la verguenza de decir que no tengo plata.
29. Me gusta acá. Es lindo para pasar verano, para tomar y para las fiestas. Es bueno sin tener que trabajar. Nadie molesta.
30. La gente aquí es envidiosa.
31. Vuelven para unas buenas vacaciones y cuando no tienen plata, se van de nuevo.
32. Ahora quiere volver a Argentina por lo que se acabó la fiesta de Chaguaya y también por lo que ya ha gastado su plata.
33. En Tarija es menos el trabajo, sólo en la casa y se paga más.
34. Aquí es trabajo más liviano, apenas de la cocina. No hay animales para ver.

# References

Aitken, S., López Estrada, S., Jennings, J. and Aguirre, L.M. (2006) Reproducing life and labor: global processes and working children in Tijuana, Mexico, *Childhood: A Global Journal of Child Research*, 13(3), 365–88.

Albornoz, O. (1993) *Education and Society in Latin America*, London: Macmillan.

Ansell, N. (2004) Secondary schooling and rural youth transitions in Lesotho and Zimbabwe, *Youth & Society*, 36(2), 183–202.

Ansell, N. and van Blerk, L. (2007) Doing and belonging: toward a more-than-representational account of young migrant identities in Lesotho and Malawi, in R. Panelli, S. Punch and E. Robson (eds) *Global Perspectives on Rural Childhood and Youth: Young Rural Lives*, London: Routledge.

Beazley, H. (2003) The construction and protection of individual and collective identities by street children and youth in Indonesia, *Children, Youth and Environments*, 13(1), retrieved 24.03.06 from <http://www.colorado.edu/journals/cye >.

Bey, M. (2003) The Mexican child: from work with the family to paid employment, *Childhood*, 10(3), 287–300.

Bynner, J. (2001) British youth transitions in comparative perspective, *Journal of Youth Studies*, 4 (1), 5–23.

Carpena-Mendez, F. (2007) 'Our lives are like a sock inside-out': children's work and youth identity in neoliberal rural Mexico, in R. Panelli, S. Punch and E. Robson (eds) *Global Perspectives on Rural Childhood and Youth: Young Rural Lives*, London: Routledge.

Chant, S. and Radcliffe, S. (1992) Migration and development: the importance of gender, in S. Chant (ed.) *Gender and Migration in Developing Countries*, London: Belhaven Press, 1–29.

Chawla, L. (2002) The effects of political and economic transformations on children: the environment, in N.H. Kaufman and I. Rizzini (eds) *Globalization and Children: Exploring Potentials for Enhancing Opportunities in the Lives of Children and Youth*, London: Kluwer Academic/Plenum Publishers.

Jeffrey, C. and McDowell, L. (2004) Youth in a comparative perspective: global change, local lives, *Youth & Society*, 36(2), 131–42.

Jones, G. and Wallace, C. (1992) *Youth, Family and Citizenship*, Buckingham: Open University Press.

Katz, C. (2004) *Growing up Global: Economic Restructuring and Children's Everyday Lives*, Minnesota, MN: University of Minnesota.

Kaufman, N.H., Rizzini, I., Wilson, K. and Bush, M. (2002) The impact of global economic, political, and social transformations on the lives of children, in N.H. Kaufman and I. Rizzini (eds) *Globalization and Children: Exploring Potentials for Enhancing Opportunities in the Lives of Children and Youth*, London: Kluwer Academic/Plenum Publishers.

Kiely, R. (1998) Introduction: globalisation, (post) modernity and the Third World, in R. Kiely and P. Marfleet (eds) *Globalisation and the Third World*, London: Routledge, 1–22.

Lacy, E. (2002) The transition to 'democracy' in Latin America, in N.H. Kaufman and I. Rizzini (eds) *Globalization and Children: Exploring Potentials for Enhancing Opportunities in the Lives of Children and Youth*, London: Kluwer Academic/Plenum Publishers.

Lee, E. (1966) A theory of migration, *Demography*, 3(1), 47–57.

Liechty, M. (1995) Media, markets and modernization: youth identities and the experience of modernity in Kathmandu, Nepal, in V. Amit-Talai and H. Wulff (eds) *Youth Cultures: A Cross-Cultural Perspective*, London: Routledge.

López Levy, M. (2004) *We are Millions: Neo-liberalism and New Forms of Political Action in Argentina*, London: Latin America Bureau.

Matthews, H., Limb, M. and Taylor, M. (2000) The 'street as thirdspace, in S. Holloway and G. Valentine (eds) *Children's Geographies: Living, Playing, Learning*, London: Routledge, 63–79.

Meinert, L. (2003) Sweet and bitter places: the politics of schoolchildren's orientation in rural Uganda, in K.F. Olwig and E. Gulløv (eds) *Children's Places: Cross-cultural Perspectives*, London: Routledge.

Pérez-Crespo, C. (1991a) *Why do People Migrate? Internal Migration and the Pattern of Capital Accumulation in Bolivia*, Binghamton, NY: Institute for Development Anthropology Working Paper No 74.

Pérez-Crespo, C. (1991b) *Migration and the Breakdown of a Peasant Economy in central Bolivia*, Binghamton, NY: Institute for Development Anthropology Working Paper No 82.

Pribilsky, J. (2001) *Nervios* and 'modern childhood': migration and shifting contexts of child life in the Ecuadorian Andes, *Childhood*, 8(2), 251–73.

Punch, S. (2001) Household division of labour: generation, gender, age, birth order and sibling composition, *Work, Employment & Society*, 15(4), 803–23.

Punch, S. (2002) Youth transitions and interdependent adult-child relations in rural Bolivia, *Journal of Rural Studies*, 18(2), 123–33.

Punch, S. (2003) Childhoods in the majority world: miniature adults or tribal children?, *Sociology*, 37(2), 277–95.

Punch, S. (2004a) The impact of primary education on school-to-work transitions for young people in rural Bolivia, *Youth & Society*, 36(2), 163–82.

Punch, S. (2004b) Scrambling through the Ethnographic Forest: Commentary on the research process, in V. Lewis, M. Kellett, C. Robinson, S. Fraser and S. Ding (eds) *The Reality of Research with Children and Young People*, London: Sage.

Punch, S. (2007) Generational power relations in rural Bolivia, in R. Panelli, S. Punch and E. Robson (eds) *Global Perspectives on Rural Childhood and Youth: Young Rural Lives*, London: Routledge.

Reboratti, C. (1976) Migración Estacional en el Noroeste Argentino y su Repercusión en la Estructura Agraria, *Demografía y Economía*, X(2), 67–88.

Reboratti, C. (1996) *Sociedad, Ambiente y Desarrollo Regional en la Alta Cuenca del Rio Bermejo*, Buenos Aires: Instituto de Geografía, Universidad de Buenos Aires.

Roberts, B. (1995) *The Making of Citizens: Cities of Peasants Revisited*, London: Arnold.

Schuerkens, U. (2005) Transnational migrations and social transformations: a theoretical perspective, *Current Sociology*, 53(4), 533–533.

Silvey, R. and Lawson, V. (1999) Placing the migrant, *Annals of the Association of American Geographers*, 89(1), 121–32.

Sklair, L. (1994) Capitalism and development in global perspective, in L. Sklair (ed.) *Capitalism and Development*, London: Routledge, 165–85.

Taracena, E. (2003) A schooling model for working children in Mexico: the case of children of Indian origin working as agricultural workers during the harvest, *Childhood*, 10(3), 301–18.

Whiteford, J. (1975) *Urbanization of Rural Proletarians: Bolivian Migrant Workers in Northwest Argentina*, Ann Arbor, MI: University Microfilms International.

# Desarrollo Integral y Fronteras/ Integral Development and Borderspaces

STUART C. AITKEN

*Department of Geography, San Diego State University, San Diego, CA, 92128, USA*
*Department of Geography and the Norwegian Centre for Child Research, Norwegian*
*University of Science and Technology, 7491 Trondheim, Norway*

A boy cries in the night, behind the wall, across the street far away a boy cries, in another city in another world, perhaps. (Carlos Drummond de Andrade; translated by Mark Strand)

Juan Tomás nodded gravely. He was elder and Kino looked to him for wisdom. 'It is hard to know', he said. 'We do know that we are cheated from birth to the overcharge on our coffins. But we survive. You have defied not the pearl buyers, but the whole structure, the whole way of life, and I am afraid for you'. (John Steinbeck, *The Pearl*, 1947, p. 54)

My hope with this essay is that I can say something about the work of young people as contained and bordered by development and modernization, how those boundaries (or, perhaps, the 'whole structure' that Steinbeck alludes to in the epigram above) are broken, and what this might have to say about the ways young people are burdened heavily and also liberated by the chimera of capitalism. I argue that borderspaces are important geographic contexts for young people to find voice, identity and integration. This is an idea of space that is not inflected only by implicitly adult imaginations but is also imbued with those characteristics of freedom, dislocation and surprise that are essential to what Massey (2005, p. 29) calls an opening up of the political. A broad under-standing of borderspaces (between children and adults, school and work, boys and girls, supermarket and parking lot, and so forth) is implied by context, and it is also a material space. The specific, material border/*frontera* for this paper comes from stories of child workers in Baja California with specific focus on supermarket packers in Tijuana, Mexico.[1] What I want to get at here—in a halting, naive and experimental way—are the ways that the machinations of borderspaces and so-called economic development are tied up with what I am calling children's *integral* (from the Spanish, meaning whole and essential) development in complex, irreducible and affective ways.

I want to raise some large theoretical questions that go beyond empirical issues that I have already considered about child labor (Aitken and Jennings, 2004; Aitken *et al.*, 2006) and globalization (Aitken, 2004; Jennings *et al.*, 2006). The corpus of this earlier work suggests that there are important connections between the local contexts of chil-dren's work and global processes, that young people for the most part enjoy the responsi-bility and discipline of work, that they are not dupes to larger political, social and economic processes that impact their labor and, importantly, that they understand their contributions as vital. I combine these issues of child labor and globalization in this essay with a theoretical focus on transformative and affective political identities that foment from representations and actions that take place in particular material spaces. I make use of Deleuzian post-structural perspectives on affect to argue that young people and their labor are a remedy and an exacerbation to the chimeral promises of economic development. I want to steer clear of any kind of answers that tie down the identities of young people, while at the same time saying something about capitalism and its labored transformations amongst young people in dynamic and rapidly transforming borderspaces.

In the arguments that follow, I look at the notion of changing borderspaces and what that means for the transformation of children's labor in terms of its representation ('I am'), its material action ('I do') and related non-representational outcomes (affects). I note that it is impossible to understand the work of children in Tijuana without considering transforma-tive border spaces in all their manifestations: from the national to the regional to the city to the body and back again.

Nigel Thrift's (2004, p. 64) recent overview of nonrepresentational theory provides something that is both concrete and relational with his suggestion that affect is 'a sense of push in the world'. The push is towards an emotion and the process is akin to what Gaston Bachelard (1969), in his *Poetics of Space*, proclaims as the 'muscular conscious-ness' of the body. The push may be a tug/movement/thrust to outrage, anger and action with witness to the exploitation of child laborers or it may be a push to spiritual enlight-enment and stillness through the beauty of child laughter in the workplace. These are pushes towards emotions that inspire the world. It is to the ways that young people experi-ence the world and are inspired through work that I want to get to with this project.

Following Deleuze (1986), affect alludes to the motion part of emotion that sloshes back and forth between perception (I-am) and action (I-do). Brian Massumi (2002, p. 22) argues that affect is about something that is not directly accessible to experience and yet is not exactly outside of experience either. Rendering an always pushing and active world, affect is understood as an embodied force rather than some invisible hand. It is that part of relations, connections, interactions and events that we feel. Importantly, it is not purely or exclusively the property of a single being or a push to a becoming. As such, affect is understood as emergent from relations between bodies. Different bodies and objects have different affects. An affect is a relation. Our emotions and feelings may be affected, or we may be propelled/pushed to take action. For Deleuze (1986, p. 65) affect 'surges in the center of indetermination' between the perceptive and the active, occupying it 'without filling it in or filling it up'. Affects are not within or without, interior or exterior, but something in between and indeterminate. They exist like a mobius-strip, simultaneously on the inside and the outside depending on our mode of encounter. Finally, affect is a movement of expression that carries stories between different levels of articulation; for example, between the embodied and the visceral, the moral and the valued, the mythic and actual.

In what follows, my first story is about regional border changes that are manifest a thousand miles south of Tijuana. Then, focusing on the stories of child workers in Tijuana, I argue that discourses of representation (a perception, an adultist designation of 'I-am') most often generated from the minority global north tend to underpin local development issues in majority global south contexts.[2] And yet, within the context of local material spaces, young people elaborate surprising ways of knowing themselves through their activities (claiming identity through material action as 'I-do'). My closing discussion focuses on the ways that young people and their labor contribute to *desarrollo integral*: the material reality of forms of 'development' that are wholly and essentially linked to place.

### Integrated Development and *La Frontera Portátil*

A thousand miles south of Tijuana, a small fish camp comprising four related families once more teeters on the brink of change. *Punta Playados* is located about half way down the Gulf of California, and is connected to Baja's Highway 1 by a 40-mile single track dirt road that is often impassible during the hurricane season.[3] Access to the camp is most often by *Pangas*, 20-foot fishing boats with powerful outboard motors (Figures 1 and 2).

The 11 children who live at the camp can trace their ancestry back six generations, to when the first Jesuit missionaries arrived in Baja California. Each family owns a boat, which the men of the community use to fish the waters of the Gulf. By mid-afternoon

**Figure 1.** Children of *Punta Playados* get ready to take panga rides to La Paz (photograph by Felipe Rickets).

**Figure 2.** *Punta Playados'* school room (photograph by Dennis Hyndman).

they take the fish to La Paz, Loreto or Santa Rosalia, where it is distributed to local restaurants catering to a tourist trade that is growing and a local population whose proportions are increasingly foreign-born (Figure 3). About 100,000 US citizens are known to live in Baja, plus an unknown number of illegal residents (Dear and Burridge, 2005). About 10% of Loreto's residents were born in the US, and since it recently became legal for foreigners to hold US title insurance on Mexican properties it is expected that this number will increase dramatically. Adjacent to Tijuana, for example, 25% of the beach community of Rosarito is now US born.

The children of *Punta Playados* learn to labor with their fathers and mothers. They catch and prepare fish, and generally help out in the community. Four weeks out of six, the community pays for a teacher from Loreto to live at the camp. During those weeks the children spend their mornings in a small cinder-block school-room. Play in the afternoon is often centered around the tidal areas or in the foothills of the *Sierra de la Giganta* that drop

**Figure 3.** Abandoned fish camp near La Paz, now used as a pristine beach for tourists willing to pay for panga access (photograph by Dennis Hyndman).

precipitously to the shore in this part of Baja California. There are few store-bought toys in the community; when the surf is gentle, two of the younger boys play with exquisitely carved models of pangas.

Although the Gulf of Mexico has supplied its inhabitants with adequate livelihoods for many generations, there is always change. The pearl fisheries that Steinbeck wrote about arose with European demand in the late nineteenth century and then disappeared in a season when disease wiped out the oyster beds. Also in the late nineteenth century, the fisherfolk of *Punta Palyados* made a good living selling fish to the miners at Santa Rosalia which, for a time, produced more copper than any other place in the world. Today, the fish stocks in the Gulf are diminishing significantly and the main economic staple is tourism. Noting the number of commercial kayak trips that paddled past his beach, one of the men of *Punta Playados* decided to start selling fish directly to tourists. He went out in the early morning to find kayakers who he then took orders from along with an indication of where they intended to camp that night. Rather than fishing all day and taking a meager catch to Loreto, he caught a specific number of lobster, parada or yellow-tale for his clients who willingly paid up to $7USD a kilo. He found that telling a few good stories about his adventures on the Gulf or advising on the best, surf-free beaches to camp upon helped fetch a better price. As the Gulf fish and lobster stalks decline, prices may increase and help sustain the community of *Punta Playados,* but at some point change is inevitable.

Some young people amongst the Gulf fish camps are turning to much more lucrative and risky practices such as drug running. Others have become seasonal migrants to the rich agricultural areas on the Pacific side of Baja near Ciudad Constituciòn, Guerrero Negro, San Quitín and Ensenada. And often young people are attracted to the economic promise of Tijuana, Tecate and Mexicali. Michael Dear and Andrew Burridge (2005, p. 6) note that California and Baja California have always had strong intertwined destinies and are presently powerhoused by the urban and economic development focused in the San Diego and Tijuana–Tecate–Mexicali metropolitan regions.

If they get established further north, these young people become magnets for the rest of their extended families. The Gulf coast boasts numerous deserted fish camps whose demise was prophesized by what Carlos Monsivaís (2003) calls *la frontera portátil.* This fluid, portable border that extends from San Diego south to *Punta Playados* and beyond; it is powered by young people who carry it with them everywhere and at all times.

The point of this beginning story is to set the stage for the notion of border fluidity. Dear and Burridge (2005, p. 5) argue that '[i]n terms of geography . . . the 'border' is something within each of us, an individual and collective mentality that is activated each time we contemplate a literal or metaphorical crossing'. Claire Fox (1999, p. 33) characterizes the border as a 'contact zone' where the terms of self-identification are relative and where centuries of complex histories, geographies and migrations produce multi-layered social networks. She points out that borders as material places, as literal crossings, have received only scant attention in cultural studies, which tend to favor it as an abstract metaphor for transnational studies rather than a concrete locale of transformative behaviors. The problem with many studies is a lack of engagement with site specific actions, because they regard the border metaphorically as a marker of hybrid or liminal subjectivities (cf. Anzaldua, 1999). The remainder of this essay is taken up with stories from supermarket workers in Tijuana. They are either recent migrants or are the sons and daughters of recent migrants. Some have roots in Baja, others come from central Mexico. I argue that these young men and women create new mental and material cartographies that proclaim what Dear and Leclerc (2003, p. 1) call collective post-border futures.

## Collective Post-border Futures

The young men and women who work as packers in Tijuana supermarkets stand out at the end of the cashiers' counters. Their un-paid industry borders the check-out stands and creates a conduit to the parking lot if you need help transferring groceries to your car. Their uniforms and hats earn them the collective nickname *Los Cerillos* (matchsticks) (Figure 4). You cannot miss them, their performance demands attention. They hope that the attention attracts a tip. They are border crossers at many levels.

I want to take some time to think about these young workers in Tijuana who border, simultaneously *Los Cerillos* (matchsticks/representations/I-am) and *Los Empacadores* (supermarket packers/I-do). I experiment with consciousness and identity from the Deluezian perspective of 'I-do' rather than 'I-am', and so I play with the doings/the workings of these young people as part of the ghoulishly indirect, and potentially liberating, discourses of a post-border/*la frontera*. I argue that it is material actions of these young people ('I-do') rather than the aesthetic of what they are labeled ('I-am') that foments change. And yet the labeling is important, I argue, because it is part of larger emotive discourses of development and capital. The young people I focus on are ghoulish in the sense that they are both the illness and the remedy enacting a space that is both irredeemably forsaken and infinitely empowering. The forsakenness comes in part from proclamations of the global minority north—through the International Labor Organization (ILO), the United Nations (UN) and the globalized rhetoric of supermarket chains, amongst others—on young people's rights and works. This is part of the 'I-am', which I see as a series of empty effects that are inappropriately structured around developmental outcomes.

Rather, I suggest a different way of looking at young people's work and identity as a series of 'I-do's' that stick and cohere as infinitely empowering because they are about the contexts that emote, engage and embed work rather than the work itself as an aesthetic. The 'I-am's' are representations of development/work in Tijuana—neat, smiling uniformed children at the end of the checkout counter—and may be thought of as striated spaces of capitalism from Deleuze and Guattari's (1988) differentiation between smooth and striated spaces. It is an overdetermined space: Euclidean, hierarchical, and orientated to the changing faces of development. It is composed of an aesthetic (*Los Cerillos*) that suggests structured, disciplined work (I-am). The work of *Los Empacadores*

**Figure 4.** *Los Cerillos* packing groceries in a Tijuana supermarket.

is also striated at certain times and in certain spaces; but it embraces another space, a smooth borderspace (literally and metaphorically) that is nomadic, folded, non-hierarchical, unorientatable and made up of free-work action of doings (I-do). *Desarrollo y fronteras*—embodied development and borderspaces—rigidify the 'I-do's' through a variety of pressures into something that may resist, may rework, may adhere to and may be enveloped by the changing faces of capitalism.

In merging with these borders, young people's identities/*desarrollo* are not randomly conceived and yet they are not predictable either. The illness-*and*-remedy borderspace finds an echo in I am-*and*-I do, a rapport with I and *non*-I, the latter a series of repetitions that presents the changing faces of capitalism and so-called development. The *non*-I is an imposition that for my purposes here becomes a striated space of fractal repetitions, of endless smiling uniformed children at the end of checkout counters. Nor is this representation fixed in its repetition or mutually exclusive as something that is striated. As I detail in a moment, '*Los Cerillos*' is a new fractal repetition in Tijuana, which supersedes earlier representations of indigenous children selling chiclets gum. Neither striated nor smooth spaces should be seen as pure or fixed categories but rather as constantly ebbing and flowing contested constructions and deconstructions of spaces (Jones, 2000, p. 44).

So, in what ways do Deleuzian notions of smooth and striated spaces help me understand the aesthetic and work of *Los Empacadores/Los Cerillos* as illness-*and*-remedy? The striated spaces are mostly the structured spaces of international development that subsumes the rhetoric of the ILO and the UN. Young people who become supermarket packers embrace this rhetoric as they agree to learn their labor in a way that is complicit with larger neoliberal agendas. It is a hierarchical space within which they shine as happy, well-dressed compliant workers whose wish is to get a full-time job at the supermarket (the *non*-I). It is a space where they aspire to 'become-the-same' as the managers of the supermarket.

Massumi (1992) argues that smooth space is favored as a territory for 'becoming-other'. This territory speaks to the problematic distinction between 'being' and 'becoming' that has warped discussions in the new social studies of children and focused them into a simplistic dichotomy between children as pre-adult becomings (becoming-the-same) and as social actors in their own right (see Gagen and Lund, this volume). Privileging the latter, the new sociology of childhood has tended to celebrate children's resourcefulness and creativity to the detriment of an analysis of wider social and spatial contexts (see Kjørholt, this volume). And it may also tend towards a reification of the child as a universal and self-serving category of existence and policy (Holloway and Valentine, 2000, p. 6). 'Becoming-the-same' is a painful recognition of norms; 'being' carries with it a huge burden of responsibility and separateness. Massumi's (1992, p. 95) concept of 'becoming-other' invites, in a post-structural Deleuzian sense, 'each contained and self-satisfied identity to be grasped outside its habitual patterns of action, from the point of view of its potential, as what it is not, and has never been, rather than what it has come to be'. It is the space of freedom, dislocation and surprise that Massey (2005) argues is required for an articulation of the political. In more spatially explicit terms, the concept draws on Deleuze and Guattari's (1987) notions of deterritorialization and reterritorialization. The former process deconstructs the individual and thus opens up new possibilities for existence (the dislocation within, and surprise behind 'I-do'), while the latter reassembles this intensity to form a new identity that is viable within the context in which it finds itself (the freedom of 'I-do').

## Encountering Development: Repetitions, Striations and Ghoulish Discourses

I will return to the smooth spaces, the deterritorialization and reterritorialization, of becoming-other at the end of this essay, but first I want to focus on what I see as the striations that create 'I-am' as an aesthetic for child workers out of the endless repetitions (the *non*-I) of so-called development in Tijuana. In what follows I talk about repetitions, civic boosterism and urban aesthetics as part of the representations that bolster Tijuana's construction of itself as a vibrant border city. I see these as part of a larger development discourse that is pervasive in global south contexts.

*Repetitions*

On my first visit to Tijuana in 1986, I was struck, like most naïve tourists, by the number of young children selling chiclets gum. At that time, the aesthetic repeated itself in a see-mingly endless fashion along Avenida Revolucion, the main tourist artery, and on many of the tributary arteries that feed the tourist district. These children, I learnt, are the sons and daughters of recent immigrants to the city. Most are from Baja, Mixteca, and other poor rural regions in SW Mexico. Very young children are often accompanied by a shawl-wrapped mother, while slightly older children persevere on their own. In the 1980s, it was rumored that a local municipal program gave chiclets' gum to indigenous street people so that their children could sell the product to tourists and thereby establish themselves as vendors rather than beggars.

Based on a study covering the period 1980 to 1997, Carmen Martínez Novo (2003, p. 250) notes that representations of these women and children from different sectors of local society (including merchants, municipal authorities, and social scientists) are either openly hostile or discriminatory through paternalistic love. Important here is not only a focus on the fluidity, historicity, and gendered aspects of ethnic and racial identities, but also an understanding of the construction of ethnicity and race in concrete political-economic contexts. In this sense, Martínez Novo's work is a consummate political economy of Tijuana's indigenous street vendors. She points out that in the 1980s and early 1990s, Tijuanese municipalities tried to project a good image of their relationship with street vendors despite fierce opposition from local merchants. This gambit emanated from the importance in Mexican politics of indigenous peoples as representatives of national identity and the poor (Martínez Novo, 2003, p. 253). Academics also challenged the views of merchants and the local press by arguing that indigenous vendors are honest workers of the tourist sector and should be awarded vending permits. This political economy is complicated by migration and by the border (see Punch, this volume, for a similar discussion of the importance of Argentina/Bolivia border to contextualizing young people's lives). A paternalistic attitude towards the vendors by Mexican academics results in arguments for using indigenous culture to preserve the 'Mexicanness' of Tijuana and thereby limiting indigenous vendors' use of the border. This bias, Martínez Novo (2003, p. 258) argues, 'reflects the fear and anxiety of the Mexican middle classes of losing control over indigenous people on the context of border contact and international migration'. The middle classes, she avers, are allowed to benefit from the border's economic and cultural opportunities, whereas indigenous people are not.

At the surface of this complicated political ecology is the stark aesthetic of the street vendors. Through the 1980s and 1990s, this aesthetic elaborates an endless repetition, which produces a particular kind of generalization—a *non*-I—about Tijuana and its place in globalization. It is repetition that is similar to the ubiquitous representation that Sue Ruddick (2003) identifies as exemplary of under-development: a child with large eyes staring at the camera. The representation begs our attention. It is a form of attention, however, that foments a problematic 'geography of care'. Large eyes staring at a camera generate a space whereby consciences in the minority global north are assuaged by sending a dollar or so a week to care for an orphan at a distance. Large staring eyes and, perhaps, a tug at your jacket as you walk down Avenida Revolucion, enable a geography of care that is taken care of with a few pesos thrown into a basket. Both are, as Bosco (this volume) suggests, distanced geographies of responsibility.

The 'children selling chiclets' gum' aesthetic is repeated endlessly through tourist areas in the majority global south and perhaps, problematically, demarcates for policy makers a certain point in a country's so-called development. Far from the US border, in tourist areas of the Equadorian Andes, for example, Kate Swanson (2005) articulates a very similar aesthetic (and cultural politics) to that of Tijuana on the 1980s (Figure 5). Her work brings attention to the differentiated ways in which modernization and globalization take shape in a marginalized region of the periphery. And, importantly, she demonstrates how children become central sites of struggle in debates over the 'proper' use of urban public space.

The important point that I want to make about Tijuana's transformation is that despite the complex political economy that Martínez Novo elaborates—with its class conflicts and indigenous sensitivities—the child vendor aesthetic is diminished significantly today. Some Non-Government Organizations (NGOs) still supply chiclets to Tijuana's street people, but the municipal focus is now elsewhere and, I argue, that elsewhere is driven to create a particular kind of aesthetic by the striations of development.

### Striations

When I first visited Tijuana, the municipality had a population of about half a million (San Diego county had a population of about two million), the Canadian/US precursor to NAFTA was just about to be enacted, and there was no *maquiladora* (literally, make-up) industries and no supermarkets. Twenty years later, Tijuana has a population of 1.8 million (San Diego County is about 3.5 million), 20% of its gross regional product

**Figure 5.** Mother and child chiclets sellers in Gringopampa, Equador (photo by Kate Swanson).

comes from maquiladora industries (with over 2000 plants) and it has 67 supermarkets. Tijuana is an extraordinary city that is growing and changing rapidly, bending and flexing to pressures from the south and to the north; from labor activism to neo-liberalism to neo-conservatism. According to Paul Ganster (2000), Tijuana's economy is one of the most dynamic in all of Mexico. He argues that, to a significant extent, this is due to its dependence on the economy of Southern California, particularly with respect to the tourism sector, the *maquiladora* industry, and the approximately thirty thousand commuter workers that live in Tijuana and work in the San Diego region. The borderspace leakages between Tijuana and San Diego are important, as are the disparities.

Also important is how actions within border cities like Tijuana become a crucial site for nation-building (Radcliffe, 1998; Paasi, 2002). Benedict Anderson (1983) famously elaborates the notion of 'imagined communities' through which national identities are produces and reproduced. Importantly these communities foment across all scales and involve government institutions, families, workplaces, news media and the public sphere. Walker (2006), through a study of the border as simultaneously cartographic, imaginary, material, social and aesthetic, argues convincingly that Tijuana is increasingly salient as an urban center in the production of Mexican national identity. And if that identity comes in part from the industry of young people as Anderson (1983) suggests, then it is important to note that the population for the state and for Tijuana is quite young. For the state, 46.5% of the population is age 19 or less; 24.2% of Tijuana's population is 14 years or less. Ganster (2000) notes that this young population provides great challenges to providing adequate educational opportunities and to creating enough jobs to employ new entrants in the job market. Much of Tijuana's growth can be accounted for by migration from Baja California and other areas of Mexico; and yet if migration into Tijuana were to cease, the population would still grow due to the large number of women in Tijuana's population that will be having children in coming years.

Forty-two of the 67 supermarkets in Tijuana engage young people as check-out assistants under the Paidimeta program, which is sponsored by the Labor Ministry, the Family Welfare System (*Desarrolla Integral de la Familia* (DIF)), and the supermarket chains Gigante (18stores), Calimax (23 stores), Ley (five stores) and Comercial Mexicana (six stores). Here is where some ghoulish discources find force. Paidimeta is an acronym that translates to a 'Program of Support for the Integral Development of Children Packers'. The Spanish word *integral* (whole, essential) complicates a series of contexts that relate to, among other things, children's relations to their families, their education and their sense of well-being. The program states explicitly that it contributes to the children's training/*formación* (i) as workers by providing a culture of work discipline, which included teaching responsible use of money, and preparing children to enter the world of work, and (ii) as family members by promoting children's dedication to study, personal achievement, family integration, and by trying to avoid children's participation in antisocial activities. For young people whose mothers and fathers learned to labor in fishing boats or in fields with family members, this latter context is hugely important. And yet, this structure elaborates institutionally controlled striations that compress work aesthetics and children's responsibilities to family life into something that not only reproduces workers but also the family system from which more good workers can be wrought. The striations find form through ghoulish discourse that emanate from the minority global north.

*Empty Effects and Ghoulish Discourses*

Elsewhere in the volume, Skelton looks closely at the United Nation's Convention on the Rights of the Child (1979) and the ways the discourses from the minority global north bleed inappropriately into the young hearts of the majority global south. Proclamations such us this merge with those of other, more insidious, international/northern institutions such as the World Bank and the International Monetary Fund to create a miasma of ghoulish discourses. Even the best intentions of institutions such as the International Labor Organization (ILO) to curb the worst of abuses towards children falter in the face of the complex relations that constitute child labor.

Mexican Federal Work Legislation, Articles 22, 23, 173 adhere to the ILO's Convention 138, which establishes 15 years of age as a minimum age for a person to carry out remunerated work, under the condition of completing basic education. The children who are part of the Paidimeta program, aged 8–14, are not paid laborers; they are considered volunteers learning a trade. As part of Paidimeta, the supermarkets' commitment is to train the children, to give them uniforms, to provide emergency medical services, and to make sure they do not participate in activities different from packing. Because the children are not official workers, they are not covered by any benefits beyond emergency health care.

Fyfe (1989, p. 4), whose works heavily influences ILO proclamations about child work, argues that child labor should be defined as work which impairs the health and development of children whereas child work constitute all work which detracts from the 'essential activities of children, namely leisure, play and education'. The ILO website maintains that 'individual children pay the highest price, but their countries suffer as well. Sacrificing young people's potential forfeits a nation's capacity to grow and develop' (ILO, 2005). As Tracey Skelton, and Anne Trine Kjörholt point out elsewhere in this volume, there is curious conflation between what is good for a child and what is good for a nation, and oftentimes both are contextualized within a very narrow and a very linear conceptualization of development. The ghoulish discourse of development conflates economic development with child development of a particular kind, a kind that has nothing to do with the whole (*intregral*) development of the child and everything to do with their *formación*/training.

According to the Paidimeta Program, store managers should take care of children packers. However, the relationship between children and managers is frequently conflicted. While children are considered as non-workers on the shop floor, managers treat them as if they were subordinated employees. *Los Empacadores* learn more than how to pack merchandise, they know the codes of products, they know where products are located, and managers take advantage of this knowledge in different ways, asking children to check prices and look for products inside the store, even when internal rules establish that the only thing children have to do is pack groceries (Figure 6). Some of the older children are given responsibilities to manage and discipline other packers; in some cases they have the authority to fire other children (see Aitken *et al.*, 2006 for examples of how these processes work in different stores). In addition, most supermarkets carry records of children packers with information about them and their families. When children do not conform to the rules established for formal workers, a disciplinary report goes in their file. We found evidence of reports on children when they laugh, run and play in the supermarket floor (Aitken *et al.*, 2006).

But the multiplicities of actual working young people are not located within these documents. To a large degree, the documents—from ILO proclamations to Paidimeta mores to disciplinary reports—are all empty effects. Of course, a filed disciplinary report is not an empty effect; it may have huge material consequences. Rather, what I am saying is that, in total, these discourses of rhetoric and documentation are striated, hierarchical fractal geometries that cannot accommodate different

**Figure 6.** *Los Empacadores* learn more than just packing groceries.

angles of encountering these young people: the emotional potency of their laughter, the fluidity of their movements. The children's flows and transgressions—their compliances and resistances—are typical of material existence, and they are the primary means through which the representations and identities of *Los Cerillos* (The I-am's) are broken down and discredited. These young people are more than uniforms that bear only a superficial resemblance to matchsticks: raising the term *Los Cerillos* as their collective identity almost always angers them. *Los Empacadores* live at the interplay of home, school and work, and as social actors they transform these different spaces, appropriating them by their social practices and in non-representational ways, and many times resisting and challenging traditional definitions of child and childhood. Their I-do's are *integral* to the larger issues of *formación* that are about transformation rather than training.

*Back to Repetition*

But the chimera of capitalism does not focus on these material contexts, at the moment it focuses on repetition and the aesthetic of *Los Cerillos*: happy, smiling bodies in clean uniforms willing to pack your grocery bags and carry them out of the clean, modern supermarket. This aesthetic (the *non*-I) displays the development and modernization of Tijuana. It is about civic pride and municipal development; a public demonstration that goes beyond the tourist areas, permeating the whole metropolitan area. And it takes me back 30 years to my first job—in the struggling economic periphery of Paisley,

**Figure 7.** Site of the first supermarket in Paisley, Scotland.

Scotland—when Tijuana occupied the exotic periphery of my grammar school geography textbook. I was a packer in the town's first supermarket, under a special governmental provision that allowed me to work while under the legal age. I learned to pack groceries (my job description) and was also press-ganged by management to price goods, compact garbage, work the cash register and, on one occasion, kill rats in the warehouse. My context as a child worker, and the context of Paisley, moves me from repetition to the suggestion of a generality (Figure 7).

I have argued elsewhere that repetition and the ways it produces stable and known developmental structures may be likened to fractal geographies (Aitken and Lukinbeal, 1997, 1998; Lukinbeal and Aitken, 1998). As such, mainstream developmental discourses repeatedly represent a progressive bargain that mirrors thoroughly known social and spatial orders. The chiclets sellers are found all over the tourist map of developing nations, the aesthetic in Tijuana and elsewhere evolves into smartly dressed packers in supermarkets. Their brown apron uniforms eerily repeating the uniform I wore 30 years ago and 6000 miles away in the periphery of the global majority north. In the same way that viewing fractal patterns of physical features like coastlines does not change with the scale of the representation, so repetition up and down a hierarchy of scales traces a neo-liberal logic of representation across a seemingly developing landscape. Taken this way, repetition is about representing change and development in a hall of mirrors and the same image recurs again and again into infinity. Like fractals within fractals, these images engage a logic of capital that is knowable and is reified continuously so that—like watching images in a hall of mirrors—I learn nothing of its constituency, only its outcome. And the outcome is quite daunting, as Guy Debord (1983, p. 40) famously points out:

> The spectacle is the moment when the commodity has attained the *total occupation* of social life. Not only is the relation to the commodity visible but it is all one sees: the world: the world one sees in its world. Modern economic production extends its dictatorship extensively and intensively. In the least industrialized places, its reign is already attested by a few star commodities and by the imperialist domination imposed by regions which are ahead in the development of productivity. . . . It is

*all* the *sold labor* of a society which globally becomes the *total commodity* for which the cycle must be continued.

Deleuze characterizes repetition in a slightly different way, and cautions that repetition is not necessarily generalization, and the two should not be conflated. To the degree that generalities are seen to be the effect of repetition, both become subsumed under the same problematic categorical heading, but Deleuze makes clear that generalities and repetition should remain distinct from one another:

> In its essence, repetition refers to a singular power which differs in kind from generality, even when, in order to appear, it takes advantage of the artificial passage from one order of generality to another. (Deleuze, 1994, p. 3)

This fundamental criticism of representation is significant, for it relates directly to my inability to recognize the multiple characters of children amongst the chiclets sellers and *Los Cerillos*. What I recognize is multiple renderings of the same narrative, the same developmental story, the same logic of neoliberalism. Steven Pile (1996, p. 223) talks about the substitution of the brutal, dark, satanic, penetrating body politic of urbanized capital by 'the cityscape as a collection of postcard scenes'. From this perspective, Pile argues, the skylines of Hong Kong, Manhattan, Singapore, Los Angeles (and maybe, in time, Paisley and Tijuana) become the perpetually acceptable, picture postcard, face of capitalism. So too, the move from an endless aesthetic of chiclets sellers to an endless aesthetic of smart, clean checkout packers are repetitions that point to development. The endless repetition and sameness becomes a generality.

And so, a new aesthetic repeats itself throughout Tijuana: *Los Cerillos* are continuously reinstituted as *non*-I's. In the 67 supermarkets from the border-end of Avenida Revolucion, through middle-class and upper-class neighborhoods in Chapultepec to low-income colonias such as Maclovio Rojas, children from 8 to 14 years (and sometimes younger) stand at the end of checkout counters packing groceries. The children are the sons and daughters of recent immigrants to the city. They wear matching aprons and caps and are the *matchsticks*. The aesthetic is found in Paisley 30 years ago. It is repeated throughout much of Mexico today, but it is in Tijuana that the government program finds its greatest success. The illness-*and*-remedy borderspace capitulates with neo-liberal coherences between the US and Mexico that produce interesting distance decay curves away from the border. The great success of Tijuana stems from Piademeta's ability to keep labor unions away from the young people. Elsewhere, and particularly in Mexico City (but also in Paisley 30 years ago), threats of involvement by organized labor immediately is countered by supermarkets pulling their support from programs that are set up to protect the welfare of children. This is manifestly ill.

## Breaking Down Striations: Smoothings and Becoming-Other

Illness-*and*-remedy. The French social scientist Michel Bonnet—despite playing a decisive role in the International Program for the Elimination of Child Labor (IPEC) on behalf of the ILO—noted in 1999 that 'one shouldn't be hypnotised by the problem of child labour, but instead should open one's eyes and ears to working children and listen to what they have to say to us' (quoted in Liebel, 2003, p. 265). The 'I-am's', infinitely repeated through striated time and space, are part of the illness of child labor and the simultaneous smooth time and space of 'I-do's' are part of the remedy. And what we hear if we listen to the Tijuanan child workers is that, for the most part, *Los Empacadores* do not see themselves as profiteers or as objects of goodwill or the concern of adults or of adult

institutions, but as independent individuals who can judge and design their lives themselves and can contribute something to society (see Aitken *et al.*, 2006 for quotes from
interviews). Their work is social and spatial, involving self-identification and dignity.

Through the larger study of Tijuana child workers, we developed new ways of thinking
about children's work as a *integral*; a multiplicity of individual actions within the context
of the Tijuana that acknowledges the benefits and risks of this work to the children, their
families and society. Striations and smoothings collide and collect, moving and transforming the shop floor, and it is a process in which *Los Empacadores* play a large part. As
visible as they are, their endless repetition as *Los Cerillos* produces what Massumi
(1992) calls 'camouflage' behind which the children work within the existing order to
ensure their own survival and that of their families and communities. Bodies-in-becoming
are 'capable of simulating the molar being assigned to them by the grid of political value
judgment' (Massumi, 1992, p. 105). This is a risky orientation, because it is all too easy to
become who you are/I am, trapped into operating entirely on hegemonic terms that is,
solely, the aesthetic of *Los Cerillos*. The trick, argues Massumi, is to throw off the
camouflage as soon as you can and still survive. This 'coming out' is never complete.
What is important is the process and the desire. This desire, I argue, is the core of a
child's *formación integral*.

## Children's *Formación Integral*

As a spatial and social practice as well as an economic practice—as *Los Empacadores*—
the children gain experience of equal and respectful relationships with other employees,
their employers and their families. They become more aware of their skills and options.
Not only does this help them to judge their family's situation better it may also open
opportunities for action. The work of *Los Empacadores* on the shop floor and at the checkout is part of a performed social space, in which the children can experience themselves
and improve their context. As such, the jobs bring self-will and dignity. And there is an
important part of their non-representational emotive space (I-do's) that cannot be overlooked. This is not just about uniforms (I-am), but also about the children's actions, movements, work and play on the shop floor as their bodies occupy and glide through, in and
around this space. As the space is filled with their laughter and energy, the children
sidle along and *come out*, throwing off their camouflage. In this space and beyond it
they are in many ways a cultural and social project, which holds a mirror up to the
society of adults and brings forth new visions and a chimera of a better life. This is,
then, not only about the child who dutifully puts on a uniform and works to better her
or himself, it is also about the subaltern who understands gravely the contribution they
make to their familial context and laughs heartedly with the society of their peers and
in the face of larger societal constructions of childhood and dignity. The dignity these children grasp lies in a uniform that only covers and does not elide a mobile, energetic and
youthful embodiment of play.

For Deleuze, the individual is really a multiplicity. The discussion of children's
becoming-other, then, signals that young people of all ages are always in motion—both
as people in their own right as well as people with specific characteristics and abilities—rather than political or economic objects that notions of 'I-am' (the 'underclass',
the 'informal sector', *Los Cerillos*) elaborate. *Los Empacadores*/I-do are about substantial
transformation, and yet the change is not conceived as the emergence of another being.
The notion of a multiplicity is a rejection of representations and generalities, of strict divisions between cause and effect, in favor of angles of encounter and non-representations.
The past, the present and the future reside in the individual. The will to power, at its best, is

characterized by *Los Empacadores* capacity for creativity rather than their desire to dominate (Deleuze, 1983).

Tijuana's child labor issues speak to the changing faces of global capitalism locally and how these aesthetics play out in the lives of young people. And this is where the story of the doings of *Los Empacadores* find power amongst the repetitions of *Los Cerillos*. Put simply, the supermarket packers in the cultural sphere of Tijuana encounter the transforming effects of everyday life, which they embrace, resist, and/or reorganize depending on the moment and the place. In the new supermarket places, a young body's performative actions is bound to perceptions and representations as well as to the peculiarities of time and space, but it is also reworking and resisting at a non-representational level and this, I argue, is where its power resides. The uniformed smiles of *Los Cerillos* are disentangled and unfolded from a generalized self-conscious aesthetic (I-am) to become the emotive work, play and frolicking of *Los Empacadores* (I-do).

## Notes

1. Published elsewhere are two empirical papers from a larger project with Silvia Lopez at El Colegio de la Frontera Norte and Joel Jennings at Cambridge University. My colleagues and I use interviews and focus groups to talk about specific issues of child labor amongst Tijuana's child-packers to the extent that I intend only a brief, general discussion of the empirical aspects of our work here (see Aitken *et al.*, 2006; Jennings *et al.*, 2006).
2. Samantha Punch (this volume and elsewhere) argues that describing the world from the perspective of global minority north and global majority south is more accurate than using anachronistic terms such as first and third world.
3. *Punta Playados* is a fictitious name. What is outlined here is an amalgamation of stories from several fish camps I visited between Mulege and La Pa, 1999–2006.

## References

Aitken, Stuart C. (2004) Placing children at the heart of globalization, in Barney Warf, Kathy Hansen and Don Janelle (eds) *World Minds: Geographical Perspectives on 100 Problems*, Norwell, MA: Kluwer Academic Publishers, 579–84.

Aitken, Stuart and Lukinbeal, Christopher (1998) Mobility, road geographies and the quagmire of terra infirma, in Steven Cohen and Ina Rae Hark (eds) *Road Movies*, Chapter 16, London: Routledge, 349–70.

Aitken, Stuart and Lukinbeal, Christopher (1997) Of heroes, fools and fisher kings: cinematic representations of street myths and hysterical males, in Nick Fyfe (ed.) *Images of the Street*, Chapter 10, London: Routledge, 141–59.

Aitken, Stuart C. and Jennings, Joel (2004) Clarity, rights and children's spaces of discipline, in Roxanna Transit (ed.) *Disciplining the Child via the Discourse of the Professions*, Springfield: Charles C. Thomas Publisher Ltd, 130–55.

Aitken, Stuart, Lopez Estrada, Silvia, Jennings, Joel and Aguirre, Lina (2006) Reproducing Life and Labor: Global Processes and Working Children in Tijuana, *Childhood*, 13(3), 365–67.

Anderson, Benedict (1983) *Imagined Commnunities*, London: Verso.

Anzaldua, Gloria (1999) *Borderlands/La Frontera: The New Mestiza*, 2nd edition, San Francisco, CA: Aunt Lute Books.

Dear, Michael and Burridge, Andrew (2005) Cultural integration and hybridization at the United-States-Mexico bordlerlands, *Cahiers do Géographie du Québec*, 39(1), 1–17.b.

Dear, Michael and Leclerc, Gustavo (2003) The postborder condition: art and urbanixms in Bajualta California, in Michael Dear and Gustavo Leclerc (eds) *Postborder City: Cultural Spacs of Bajalta California*, New York and London: Routledge.

Debord, Guy (1983/2000) *Society of the Spectacle*, Detroit, MI: Black and Red.

Deleuze, Gilles (1983) *Nietzsche and Philosophy*, New York: Columbia University Press.

Deleuze, Gilles (1986) *Cinema 1: The Movement-Image. Trans.*, Hugh Tomlinson and Barbara Habberjam, London: The Athlone Press.

Deleuze, Gilles (1994) *Difference and Repetition*, New York: Columbia University Press.

Deleuze, Gilles and Guattari, Félix (1983) *Anti-Oedipus: Capitalism and Schizophrenia*, Minneapolis, MN: University of Minnesota Press.

Deleuze, Gilles and Guattari, Félix (1987) *A Thousand Plateaus: Capitalism and Schizophrenia*, London: The Athlone Press.

Drummond de Andrade, Carlos (2004) Boy crying in the night, in Fred Moramarco and Al Zolynas (eds) *The Poetry of Men's Lives*, Atlanta, GA: University of Georgia Press, 20.

Fox, Claire (1999) *The Fence and the River, Cultural Politics at the US-Mexico Border*, Minneapolis, MN: University of Minnesota Press.

Fyfe, A. (1993) *Child Labor: A Guide to Project Design*, Geneva: ILO.

Ganster, Paul (2000) Introduction to Tijuana and its Region. http://www-rohan.sdsu.edu/~/irsc/public.htm/.

Holloway, Sarah and Valentine, Gill (2000) Children's geographies and the new social studies of childhood, in S. Holloway and G. Valentine (eds) *Children's Geographies: Playing, Living, Learning*, New York and London: Routledge, 1–26.

Jones, Owain (2000) Melting geography: purity, disorder, childhood and space, in S. Holloway and G. Valentine (eds) *Children's Geographies: Playing, Living, Learning*, New York and London: Routledge, 29–47.

Jennings, Joel, Lopez Estrada, Silvia and Aitken, Stuart (2006) Learning and earning: relational scales of children's work. *Area*, 38(3), 231–39.

Kalachek, E. (1969) Determinants of teenage unemployment, *Journal of Human Resourcses*, 4(1), 3–21.

Kalachek, E. (1973) *Labor Markets and Unemployment*, Belmont, CA: Wadsworth Publishing Co.

Liebel, Manfred (2003) Working children as social subjects: the contribution of working children's organizations to social transformations, *Childhood*, 10(3), 265–85.

Lukinbeal, Chris and Aitken, Stuart (1998) Sex, violence and the weather: male hysteria, scale and the fractal geographies of patriarchy, in Heidi Nast and Steve Pile (eds) *Places through the Body*, London and New York: Routledge, 356–80.

Martínez Novo, Carmen (2003) The 'culture' of exclusion: representations of indigenous women street vendors in Tijuana, Mexico, *Bulletin of Latin American Research*, 22, 249–68.

Massey, Doreen (2005) *For Space*, London: Sage Publications.

Massumi, Brian (1992) *A User's Guide to Capitalism and Schizophrenia*, Cambridge, MA: MIT Press.

Paasi, A. (1991) Deconstructing regions: notes on scales of spatial life, *Environment and Planning A*, 23, 239–56.

Pile, Steve (1996) *The Body and the City*, New York and London: Routledge.

Radcliffe, Sarah (1998) Frontiers and popular nationhood: geographies of Identity in the 1995 Ecuador-Peru border dispute, *Political Geography*, 17(3), 273–93.

Ruddick, Susan (2003) The politics of aging: globalization and the restructuring of youth and childhood, *Antipode*, 35(2), 334–62.

Steinbeck, John (1947, 1992 edition) *The Pearl*, New York: Penguin Books.

Swanson, Kate (2005) Begging for dollars in Gringopampa: geographies of gender, race, ethnicity and childhood in the Ecuadorian Andes, PhD Thesis, University of Toronto.

Tarcanera, Elvia (2003) A schooling model for working children in Mexico, *Childhood*, 10(3), 301–18.

Taracena, Elvira and Tavera, Maria-Luisa (2000) Stigmatization versus identity: Child street-workers in Mexico, B. Schlemmer (ed.) *The Exploited Child*, London: Zed Books Ltd, 93–105.

Thrift, Nigel (2004) Intensities of feeling: towards a spatial politics of affect, *Geografiska Annaler*, 86, 57–78.

Walker, Margath (2006) PhD thesis. University of Kentucky.

# At the Interface of Development Studies and Child Research: Rethinking the Participating Child

RAGNHILD LUND

*Department of Geography, Norwegian University of Science and Technology (NTNU), NO-7491 Trondheim, Norway*

## Introduction

During recent months I have had the chance to spend time in several African and Asian countries, amongst others Sri Lanka, where I have travelled to and from for the last 30 years, and another which is fairly new to me, namely China. It has been both intriguing and thought-provoking to learn how substantially different the livelihoods of children are in different Asian countries.

I spent a month in Wuxi and Shanghai, interviewing workers in Norwegian companies. Through talking to female industrial workers, especially, I gained insights into some of the impacts of the new industrial economy on Chinese family life and local communities. A major bottleneck in the new market economy is lack of child care facilities. Young women work under the new Chinese regime with globalised production in primarily foreign-owned companies. The children can no longer come with their mothers to be taken care of in the communal crèche. The latter no longer function in the new economy of China as such services are now being privatised and are rare. The majority of workers cannot afford to pay for private day care. Instead they send their child to the grandparents who retire early if necessary to look after their grandchildren, sometimes in the hometown far away from the workplace of the young parents. The one-child policy still prevails. So, basically, the grandparents look after the one, very precious child, who has been 'bestowed' on the family. The child is pampered, idealised and spoon-fed in every kind of way, especially if it is a boy. The girl child is considered less precious, as she can do less, it is thought, to provide for the future of her family. Officially too, children are being seen as the future of New China; they are being portrayed as fortunate, well-fed, well-dressed, and well-behaved. The child does not have to worry—China is looking after its precious young.

This situation is tremendously different compared to the situation in Sri Lanka, a country deeply entrenched in poverty, conflict and war, and most recently hit by a tsunami (Boxing Day, 2004). The poorest, most marginalised people of the Sri Lankan society have suffered the most due to the tsunami. Fishermen and refugees settled along Sri Lankan shores lost family members, their homes, even their whole local communities. When the waves hit, mothers wearing their traditional lungi or saree could not swim or save their children. The majority of the tsunami victims were therefore women, children and the elderly. In other cases, children were left behind when their mothers and fathers drowned. In the north and east, the tsunami victims were people who had already lost their homes, including women who had lost their husbands and children due to the war. To date, *c*. 35,000 people have been killed due to the tsunami, and approximately 1 million people have been displaced; 1030 children are reported as orphans with no parents or relatives left. The tragedy is without precedent in Sri Lankan history.

These two different experiences made me realise again that there is no such thing as the Global South. Situations in the South (between and within states) may vary as much as situations in the Global North and between the North and the South. While my different experiences provided an eye-opener to the varied situations of the children in Asia alone, it also provided an eye-opener to what constitutes childhood and how childhood is constructed by local places and cultures, nature as well as geopolitics. In fact, childhood is constructed in a complex matrix of these factors, and where unintended and intended events play a significant role in the formation of people's lives and livelihoods.

In the following, contingent representations in development and child research will be explored to see if they together may lead to a new gaze on childhood and how the agency of children may be analysed. This chapter examines what is perceived as *participation* and the *participating child* in both research traditions. Both discussions are problematic in their failure to address adequately wider contexts in which adults and children live. However, the meaning and significance of the term has changed over time and pertains to various scales and levels of abstraction. While initially the term participation was seen as a means to change practices to serve basic needs at individual and community levels, participation now is increasingly seen as a desired outcome of development to obtain individual and collective rights. In addition to the practice-oriented scope, participation has become part of a global discourse of development. While *globalisation* is seen as a structural force shaping children's autonomy on how they may perform and how they may be mobilised by self and others, participation has been seen as a way to encounter global forces and how children may mobilise in this process.

## Participation

Participation has become a major feature of contemporary approaches to poverty reduction and social change. Organisations ranging from small community-based organisations

(CBOs) and non-governmental organisations (NGOs) to international agencies have recognised the potential for increased participation in policy processes. Participation is generally voluntary, but may not necessarily be an involvement in people's own develop-ment. People may participate in projects without influencing them, for instance (they are beneficiaries of development projects, not stakeholders), while in other cases they may actively involve in changing their lives (e.g., stakeholders in unions).

Hence, participation may take different forms, ranging from the passive to the active, even interactive or self-mobilising to change institutions or systems. It may be seen as a way to sensitise people on issues that need action, as a way to actively take part in development processes, as a way to foster dialogue between different stakeholders and to influence decision-making, and as a right to influence policy making and processes. Various structural forces are at play and impacting on the levels and types of participation, such as parents, families and peers, and internal or external, local and global institutions (Williams, 2004; Ansell, 2005). In addition, participation has to take place at the most con-venient time and place (Cornwall, 2002).

At this juncture it may be legitimate to ask whether the concept participation is over-stretched as an analytical and practical term because it encompasses multiple layers of meanings and practices, and it refers to different scales and types. However, the focus on participation is still important because it signifies ways to sensitise, change and empower people, it may provide insights into processes of inclusion as well as exclusion, and it provides insights into how children are perceived locally and globally (Mohanty *et al.*, 2006). Also, it may lead to an understanding of the processes of margin-alisation, even exploitation, of children.

## The Quest for People-centred Development

The quest for people-centred development approaches has come from various positions in the development field. It was first voiced in the 1970s and gained momentum at the Cocoyoc conference in Mexico in 1974 (Nerfin, 1977). It was geared towards the satisfac-tion of needs, and endogenous and self-reliant development in harmony with the environ-ment. In the mid-1980s, Booth (1985) and others also argued for alternative development perspectives that are inclusive, participatory, and sensitive to diversity and difference. In the 1990s, emphasis was put on processes that went beyond the satisfaction of needs to empowerment of individuals and households through their involvement in socially and politically relevant actions, and was coined *Alternative Development* (Friedmann, 1992; Lund, 1994; Pieterse, 2001).

Alternative Development theory is oriented towards understanding agency. It basically refers to a bundle of normative development theories which deal with people-centred development. Alternative Development is not new. Inspired by feminism, the Black Movement in the US, deep ecology, and others, it developed in the 1970s as a quest for 'another development', based on indigenous values, self-reliance, fulfilment of basic needs, dignity, and the redistribution of goods and services (Nerfin, 1977). It was not until the early 1990s that this approached was criticised for being too focused on fulfilling needs rather than identifying ways to empower the poor and marginalised (Friedmann, 1992; IFAD, 1992; Lund, 1993a). Part of this critique came from feminists who questioned how men and women could sensitise themselves and act against oppressive structural forces (Lund, 1993a,b).

However, the new critique had little concern with individual human actors, such as chil-dren, or their voice and their role in society. Children as individuals and groups were largely invisible.

Another critique of development studies and the development concept is grounded in postmodern and post-colonial theory that refers to the 'development project' as basically *Eurocentric* (Mehmet, 1995; Escobar, 1997). From this vantage point it becomes important to explore the intersections of various forms of power in relation to participation, vulnerability, class, and gender. There is also a need to contextualise, look at variety, recognise other knowledge systems, and seek alternative 'truths' (Tuwai Smith, 1999). In this perspective, development theory, its key concepts and analytical tools are seen as basically *Eurocentric* and hegemonic constructions which need to be deconstructed in order to understand what social change is all about (Escobar, 1997). Participation may be seen as one such Eurocentric concept; it may not be appropriately contextualised.

These theories echo the quest to conceptualise childhood as a social construction that has to be analysed with respect to the culture and society to which the child belongs and the prevailing norms and structures that impact on a child's life and performance (Qvortrup, 1997; Kjørholt and Qvotrup, 2000; Kjørholt, 2001). Childhood is perceived as a distinct category in society. It is defined by its relations within our generational structured world. Individuals are constructed as children, while others are constructed as adults. The members of each category are ascribed as having unique competences for their category's activities and identities (Moono, 2006).

The constructionist view also acknowledges that children are understood in different ways in time and place. The role and structure of the family vary greatly in character from one type of society to the other, for example. While the nuclear family pattern dominates in the West, in Africa and Asia the extended family pattern is important. The roles performed by biological parents and children in the West are often different to those in other cultures. Nurturing and care may be different or changing. In much of Africa and Asia, children represent lineage continuity as well as the material survival of families and communities. They are expected to provide labour and support to the older generation, the sick and needy. Most children grow up in peasant households with distinct responsibilities, where the division of labour varies according to age and gender. Sometimes the children spend more time on these duties than going to school (Ansell, 2005). My own experience from studying orphans in Ghana shows that children may have extensive responsibilities in the household and local communities (Lund and Agyei-Mensah, 2007). Hence, the contributions of children are valued differently in different societies, and this dramatically changes human relations in situations of vulnerability.

Likewise, with respect to participation, there is a need to deconstruct knowledge about how child and childhood are perceived and constructed and how we can analyse how children *capacitate* (participate actively and for a change) in building on their local capabilities. In doing this, it is necessary to know more about the role of agency in processes of social change.

**The Critique of Participation in Development Theory**

Like the broader framework of Alternative Development, participatory approaches have been criticised for being nothing more than a fashionable rhetoric, which promises empowerment and appropriate development. Cooke and Kothari (2001) show how participatory development has a long history in development studies. It emerged as a critique of top-down development approaches which were considered inefficient and donor driven. It claimed that people were central to development and encouraged involvement where one could build on 'local' people's knowledge, perspectives and priorities. It also led to the development of a wide range of methodological tools which were used by NGOs,

CBOs and other development agencies. What more recent critiques have shown, however, is that the weaknesses of participatory approaches lie both in the methodology and techniques, as well as with the politics of the discourse. Cooke and Kothari (2001) claim that participation is a seductive claim which has a lot of pitfalls and limitations, as being too naïve, not culturally sensitive, and often reinforces instead of overthrowing existing inequalities. Others direct their critique towards the failure to engage with issues of governance, obsession with the local level and an insufficient understanding of how power operates and is constituted (Cornwall, 2002, Mohanty *et al.*, 2006). For all these reasons, it is argued that participatory approaches commonly fail to achieve meaningful social change for the stakeholders involved.

Alternatives to participatory approaches have come from two different streams of thought: by Norman Long who makes a quest for actor-oriented approaches in development studies (Long and Long, 1992; Long, 2001), and by Amartya Sen and others who focus on building human capabilities (Sen, 1981; Dreze and Sen, 1989; Nussbaum, 2000).

An actor-oriented perspective implies that peoples' ways of coping and their choices are determined by the societal and spatial contexts they live within. Actor-oriented approach:

> assumes that variations in organisational forms and cultural patterns are to a large extent the outcome of the different ways in which social actors organisationally and cognitively deal with problematic situations and accommodate themselves to the interest and design of living of others. Inherent in this concept of social actor is the notion of the human being as an active subject with the capacity to process social experience and to invent new ways of coping with life under extreme coercion. This holds good whether the actor is deemed 'powerful' or 'powerless'. Within the limits of their socio-cultural context the men and women attempt to solve problems, learn how to intervene in social events around them and monitor their own actions as well as observing how others react to their behaviour. (Long and Long, 1992, pp. 222–3)

From the quotation by Long and Long (1992) it may be learned that a major element of an actor-oriented approach is the emphasis put on power, choice and context. As children are generally perceived as powerless and without voice in most cultures, the actor-oriented approach provides a way to learn what constitutes a child actor and what is his/her potential for participation and change. One way of doing this is to carefully assess the children's assets (knowledge, material and immaterial resources, their creativity, their ability to organise collectively) in a given place and time, given the local, cultural and economic structural forces at play. In the case examples given below, emphasis is given on identifying assets and structural forces in order to analyse the children's abilities to cope and strategise in different situations.

The capabilities approach also emerged on the development agenda as a response to the failure of conventional development theory and practice to include the poor and to provide them with essential services that could help to meet their essential needs (drinking water, sanitation, public health, transportation). It also expanded the idea to include needs considered essential to a 'decent human life', such as shelter and clothing, but also the right and ability to take part in the community and achieve self-respect (Kabeer, 2003). These ideas are, in a way, informed by the notion that every human being, including children, has a right to live their life in dignity and freedom, namely that development should contribute to people's capabilities to achieve the lives they want (Nussbaum, 2000).

Sen's basic idea (Sen, 1981; Dreze and Sen, 1989) is that it is the capability of every individual human being, not functioning, that should be the appropriate goal for development and political achievement. Each individual's perceived well-being should be the basis for social choice. Since capabilities are not about what people can 'choose' but what they are able to achieve, they depend partly on personal circumstances and partly on social constraints (Kabeer, 2003). Furthermore, the capabilities may be seen as the foundation on which people may be able to empower. In general, it may be argued that the major determinant of children's participation is the role of adults. However, also important is the significance of the political, cultural and historical contexts in which the children and their parents live and the community they belong to. Against this background, we may search whether children are allowed freedom of choice, given their assets, age and gender. Assessing the choices that are made by the children themselves, the circumstances and constraints (social, cultural, economic) to children's participation may provide insights into their ability to change and empower. In the case examples, the term 'capacitate' is used with reference to children's ability to change their circumstances and to empower.

## Participation and Child Research

Attitudes towards children and young people and their participation have also been changing over the past decade. Putting children and young people's interests and opinions firmly onto the international agenda has been described as the new challenge for social development, and the concept and practise of children's participation have become increasingly established and accepted by organisations and governments around the world (Save the Children, 1995; Williams, 2004). This marks a shift from considering children as marginal and passive recipients of social change to the promotion of children as development targets for themselves to enhance their child rights.

In child research environments, participation implies listening to children, giving them space to articulate their own concerns, and to enable children to participate in planning, monitoring and evaluation of activities within and outside the family sphere. It is about giving children a voice and a role in decision making, and it is about seeing the participation of children in relation to older people's control and power (Hart, 1992; Van Beer, 1995; Mwale, 2001; Williams, 2004; Winsvold, 2004). In fact, there is a growing view that children are competent social actors, and if mobilised, may extend to children the rights to freedom and self-determination (Archard, 1993, in Ansell, 2005). However, in real life situations, there seem to be big differences between various countries; in their culture, within the educational system and in their 'tradition' of listening to children.

The UN Convention on the Rights of the Child (CRC) encompasses rights relating to protection, provision and, most controversially, participation. The CRC conceives of participation in two ways; possibilities for children to engage with the world around them and opportunities to have a voice in more formal decision-making processes (Stephens, 1994, in Ansell, 2005). As children are generally deprived of the possibility to participate and are silenced and made invisible by the attitudes and practices of adult society, participation is seen as a way of enhancing children's control over their own lives (Roche, 1999 and Archand, 1993, both cited in Ansell, 2005). Therefore, children who have been mobilised through participation are increasingly perceived as producers of knowledge about children and childhood because they contest established traditional attitudes and cultural values about children in local, national and international contexts. This indicates primarily positive attitudes towards the ability of children to capacitate and contribute meaningfully to social change and development.

However, children's participation is not yet universally accepted. Ansell (2005) lists the following arguments in favour of participation, which may be supported by my own observations: First, children have seldom been expected to contribute their views. They have been considered powerless and inarticulate. However, these characteristics also hold true for adults, and adults have often been poor interpreters of children's lives. Therefore, children's participation is important. Second, children are often more competent than is assumed. Even small children can make important decisions; they can negotiate and they have often wide responsibilities. This is in evidence in Africa, for instance, where due to the AIDs pandemic child-headed households manage to create a living for their siblings and elderly family members (see also Lund and Agyei-Mensah, 2007). Another example is street children who provide for their own existence. Third, children are often key stakeholders with relevant experience in matters relating to their lives; for example, children who collect firewood to sell in neighbouring towns (common in Sri Lanka). Fourth, participation may constitute an education for active citizenship (cf. Winsvold's study below). In fact, it can be deduced that children always have been important actors in change.

As the previous sections have shown, the change from viewing children as beneficiaries to viewing them as active agents has been widely debated both within and outside child development circles. On the one hand, it is generally believed that children have service-based needs rather than rights, especially because children will benefit from policies designed for adults anyway. On the other hand, children will always only be an add-on to development programmes if they cannot participate and influence decision-making processes. Through rights-based approaches, participation takes on a new dimension and helps to account for the range of attitudes and approaches to children's contributions in society. This new discourse on rights echoes the wider shift in the development discourse and practice, particularly as articulated within the alternative development framework.

However, the rules and practices of the wider society may not provide opportunities for meaningful participation by children. In the following, I will present some examples of what children can achieve and the way they participate in different social circumstances. I will ask what are the social and cultural options and constraints to children's participation—by parents/peers, the local community and state—and how do they relate to wider, global social and economic forces.

## Participation as a Means to Empower

In spite of recent critique, development thinking on participation has enormous support among NGOs, CBOs and practitioners working with special targeted groups, such as women and children. Participation is also introduced as the third 'P', alongside protection and provision in the child research and child rights environment globally (Skelton, this volume). It represents a major way to incorporate the role of agency and people's aspirations through activities such as community efforts, campaigning and lobbying. With respect to children as beneficiaries of such projects, activities may be child initiated, such as the child union example (presented below), or initiated by adults, such as the Zambian and Ghanaian examples given below.

As mentioned, participation also ranges from the receiving end, i.e., being passive or mere tokenism, to various degrees of participation, ranging from being a beneficiary to an active participating actor, where decisions are taken jointly with adults (CIDA, 1992; Hart, 1992). The situation of children in new China may illustrate how this range prevails within the same country. Little change in the situation of children is reported from rural areas. Here children are still active contributors to the rural economy (helping hands on the farm and at the local markets). However, children's lives in the rapidly changing towns and cities are transforming fundamentally, from being active participants in the local community to passive recipients of social and cultural change, sometimes with quite negative outcomes. In fact, several child experts report on misconduct and irrational behaviour of Chinese children because they are the only child, they are under-stimulated, and they are isolated because they live in city environments and are alienated from natural social surroundings (UNICEF, 2004).

A study by Lund and Agyei-Mensah (2007) shows what happens to children's participation when their parents are infected or passed away due to HIV/AIDS and the traditional safety net operates to solve the problems of increasing orphanhood. In Kroboland, Ghana, the traditional Queen Mothers and private caregivers (generally elderly women) look after the orphans by finding homes for them, by providing primary school education and vocational training, health awareness and cultural knowledge through handicraft production, dance lessons and sport activities. While analysing the role of the work of the *Manya Krobo Queen Mothers' Association* (QMA) and other caregivers, the perceptions of some of the orphans were identified. Children described how it is to be an orphan, what kind of work do they do at home, what were their dreams and hopes for the future, and what they knew about HIV/AIDS. It was found, that even though the QMA serves as a good role model for how to tackle the orphanhood problem, it faces limitations as to how much it can do. The effectiveness of its support depended on the contribution of the wider Ghanaian society to make an impact. The children, continue to participate under deprived and vulnerable situations, although they show great strength in managing households and take important decisions. Box 1 gives the story of one of the girls we met during fieldwork in August 2005.

## Box 1.

Debbie, 13, was brought to the school by her grandmother three years ago. She lives with her. Her mother is dead and she never knew who was her father. She told us that she is very lonely, because she has so much to do and she has no time for friends. Her future dream is to become a nurse to assist people in need like her grandmother.

Debbie told us that she gets up early morning, helps her grandmother out of bed and give her something to eat. Then she goes to school, sometimes she is late. After school she does the marketing and cleans the house. She has to do everything herself and has nobody to turn to in times of problems:

*(I) When you have problems, who do tell?*

*(R) I don't tell anybody*

*(I) Why?*

*(R) We don't have any mother and so my grandmother does not do anything (work) so if I am sick I don't tell anybody*

*(I) But do you ever tell the queen mothers?*

*(R) Yes*

*(I) But what if you need money? When you don't have money and you need money, who do you tell?*

*(R) My mother was attending a church before she died. If I go to the church they will give me five thousand cedis* (about fifty-five US cents)

*(I) Oh yes, I forgot to ask you. You didn't tell me where your father is?*

*(R) I didn't know my father.*

*(I) Was it only your mother you knew?*

*(R) Yes*

*(I) But do you visit your mother's relatives, your mother's brothers and sisters?*

*(R) They are not staying here, they are at Asesewa and I don't have money to there so I don't visit them and I stay here.*

Lund and Agyei-Mensah, fieldnotes, 2005

In other cases, children are deprived of their potential ability to capacitate. This may happen when a third party intervenes to help poor and destitute children (Box 2).

**Box 2.**

The study of orphans and vulnerable children by Mwene (2006) investigates World Vision International in Zambia, which is an NGO '*undertaking child focussed and community based development in all nine provinces of the country. Based on a regional approach and emphasising devolution of power to the poor, the organisation reaches about one fifth of the country's ten million people and targets those that need development the most: the poorest. Its development interventions are said to be directed at the poor's basic needs and are empowering because they involve the poor in providing for their own needs. Through this process the community is further said to be capacitated not only to meet their needs now but even beyond NGO support . . . the organisation was founded on the principle of sponsoring children in a given area for purposes of undertaking designated community development projects whose benefits are supposed to spill over to the children*' (Mwene, 2006, v and 90). In addition, more direct support is provided to children, such as medical support, food, clothing, shelter, and school fees. It was found that individual participation by sponsored children was limited to their availability during enrolment and also during ongoing child information update sessions. Beyond this, the children were passive recipients of charitable NGO support, and itemised under the kind of activities undertaken. Children were considered as too young to make contributions of their own and their interests were thought to be best represented by the adults under whose care they fell. Hence, children's provision and protection rights are emphasised over participation. However, children's participation in household chores were seen as facilitating the complete functioning of community life and hence participating in their own way. In this way, they were perceived as constructively involved in the prevailing system of production.

Mwene (2006) found that this process of participation is making a difference for the children involved but not really changing the local community to the benefit of poor and destitute children. Another study on NGOs and children in Zambia by Mwale (2001) is an assessment of a project for street children in Lusaka. The NGO runs a hostel for street children and they receive free food as well as some schooling. The children are being reintegrated into the local community through work as NGO interns, trainees and local community workers. Although a fairly successful project with respect to the participation of the children, the regime of the NGO was so rigid that the number of children returning back to the street was fairly high. Both cases show that participation has increased children's capacities but has not managed to mobilise them adequately in cases where adults have defined the needs of children.

The study by Winsvold (2004), on the other hand, is an example of child-initiated participation (Box 3). Her study of child unions investigates working children's participation, mobilisation and agency in Karnataka in India. While child labour is generally perceived as exploitative and should be abolished by strict legal regulation, this study argues that such measurements will not stop the exploitation of children but will only make it invisible, and therefore more dangerous than before. Some working children even claim their right to work, stating that work is their best option to survive and may even be more beneficial than school.

Winsvold found that the process of child participation and membership in unions was one realistic way of creating more social space for children to improve their situation. Children with different backgrounds gathered around common goals and managed to highlight common issues; they questioned power relations and created a common identity. Mobilisation was seen as the only way for children to gain recognition and respect for who they were, and what they can do and achieve.

**Box 3.**

To overcome their vulnerability and to be able to capacitate, such children have formed their own community-based organisations (CBOs), 'working children's unions'. In India there are nine child unions and they have founded their own national union, the National Movement of Working Children in India. Meetings where progress, experiences and strategies are shared take place regularly, and on several occasions they have questioned national and international institutions by *'arguing that children have enough experience, knowledge and understanding to handle complex issues and by claiming that their work could be of positive value ... [that] should be considered as a right'* (Winsvold, 2004, p. 279).

An in-depth study of one working children's union, Bhima Sangha, was investigated. It was found that boys and girls do different types of work; they work both at home and in factories and in the agricultural sector. Their earnings are lower than the minimum wages, and their pay never corresponds to the time spent on work. Most live children with their parents or other caretakers in slum areas and they claim various reasons for working, such as parents who are ill, alcoholics or poor, the need for dowry payments, or dropouts from schools (failed exams, harassment, failure to pay school fees). Their basic rationale for joining the unions is the wish to improve their lives, work with others, and acquire new knowledge. Still, the children had limited knowledge on their child rights and were dependent on the good will of the various adults/peers that supported them: *'None of them knew anything about "children's rights" before joining the Sangha. At the same time they emphasised that the "child rights" were just empty words meaning nothing unless they were recognised by the people in the surroundings. To get recognition and respect for who they are and what they do and achieve, and to get an understanding of their situation, are very important factors for the mobilised children'* (Winsvold, 2004, p. 280).

In these examples participation is seen as a liberating force; children can capacitate and thus benefit from the projects. Both the interventions by the NGO and the mobilisation by the CBO do a lot of good, but the impacts of their work are limited and fragmented. In the NGO example, children are perceived as beneficiaries and participation is used as a method of integration, sharing and competence building. While the NGOs serve the needs of the children, they do not contribute to changing their situation in society or mobilising them to overcome their vulnerable situation. The CBO, however, enables children to capacitate. This example shows how children who are active project participants become stakeholders who can contribute to social change. However, both types of approach show that participation is of limited value in cases where mobilising around children's rights is not accompanied by wider political reforms, as well as cultural changes that may improve the situations in which children live. Mobilisation thus unavoidably depends on various structural forces embedded in local and global politics.

Hence, although participation is perceived as part of a global discourse and action (CRC), it seems to have had limited empirical and practical significance. At the local level, participatory approaches have often focused on beneficiaries, not individual actors of social change. Participation has been seen as a methodology or means and, to a very limited extent, as an end or empowering force. Another weakness has been the inability to distinguish between different cultural and contextual circumstances. Participation has not been seen as a structural force that must be appropriate in time and place. As will be seen below, participation of children is also part of a global-local nexus and leads to increasing vulnerability for poor children, or in the worst and most marginalised situation it may imply exploitation.

## Participation, the Child and Geographies of Globalisation

Major theoretical discourses on globalisation have focused on its driving forces, such as the free flow of capital, goods, resources, technology, and services. In the search for understanding what cultural and social processes are at work, and what implications these may have at different scales, researchers have postulated various hypotheses: about an increasing economic integration and a new division of labour, homogenisation (we become more alike) and new patterns of differentiation, 'time-space compression' (Harvey, 1989; Robertsen, 1992; Waters, 1995), and 'time-space distanciation' (Giddens, 1989; Inda and Rosaldo, 2002).

Giddens (1989) very early stressed the interdependence of local and global events and states that local transformation is as much a part of globalisation as the lateral extension of social connections through time and space. Societies transform differently, and it becomes pertinent to understand that non-Western local traditions are also historical and dynamic. According to Appadurai (2001), the global situation is interactive and not one-sidedly dominated by an 'Americanisation' of the rest of the world. Rather, globalising and localising processes impact each other. Appadurai stresses the cultural dimension of these processes as ideas of situated differences that are related to phenomena that are local, embodied and significant to the persons involved. This argument may be extended to include children as enabling, but often marginalised, disempowered and exploited actors in certain social contexts. Below, two examples show how economic restructuring/globalisation determines children's participation in different contexts and situations, ranging from marginalisation to exploitation.

*Children's Participation and the Global/Local Nexus—The Role of the Sri Lankan Diaspora to Rural Sri Lanka*

Over a period of 30 years I have been visiting the Mahaweli project area in Sri Lanka, where the largest colonisation project in the country was implemented between the late 1970s and the mid-1980s. This project represents a major effort to reconstruct the rural economy of the country. Funded by the World Bank and other major multilateral and bilateral donors, it may be seen and analysed as the first structural adjustment programme of the country, which was aimed at solving some of the country's economic, demographic and settlement problems. It involved irrigation of 1659 km$^2$ of land (to increase productivity of the land for cash crop production), resettlement of more than 140,000 families, and the provision of power, infrastructure and other services (schools, hospitals, markets) which serve these areas. Each family was allocated 2.5 acres of irrigated fields plus 0.5 acres of home/garden land. Settlements were formed according to a hierarchical pattern of hamlet, villages and townships.

Various studies over the years, including my own (Lund, 1981, 1983a,b, 1993a,b, 1994; Mueller and Hettige, 1995), have found that the Mahaweli area has developed into an economically highly differentiated society. Many settlers have succeeded in agriculture or in businesses, while others have failed due to problems of marketing and price fluctuations, land and water quality, production problems, lack of off-farm employment or personal misfortune. In 1991, for instance, a large section of the population could not produce cost-efficiently and became marginalised and economically poor (Lund, 1993b).

Furthermore, the division of labour in the Mahaweli became blurred (Lund, 1981). First, there was no longer a distinct definition and allocation of workloads between men and women, but also no consistent views on how children could participate in daily chores at home and in the field. During the pioneer years of settlement, for instance, husbands and wives together with the help of their children shared the workloads equally between themselves, and not according to the traditional division of labour between the genders and age groups. As there had also been a transition towards a nucleated family pattern, and no longer an extended family pattern as in the traditional communities, the pioneers were faced with quite harsh working conditions, working long days out in the field, and often with the result that small children remained alone at home for large parts of the day. There were no crèches or babysitters available. In the process, the children had missed the opportunity of being part of a larger family unit or part of a close-knit village community and were more isolated than before. In fact, the situation of children was very little debated in Sri Lanka at the time, and particularly with respect to the situation in the settlement schemes.

When I revisited the area in 2004, and could observe economic improvement, it turned out that the new revival of the economy is entirely due to the new global economy and the ongoing war. Sons and daughters of the pioneer settlers are working as soldiers, industrial workers or workers/maids in the Middle East. The remittances that children send to their parents have enabled the parents to build a new house and to avoid economic poverty. It could be argued that while the pioneers worked the land, and with very simple means created a home during the early years of settlement, the children's obligations towards their parents have enabled them to avoid the poverty trap. The way the second generation of settlers strategises is thus interlinked with globalisation and making a life in both the local village and in the Sri Lankan diaspora. However, people work in the Sri Lankan diaspora without their dependants. Grandparents and young children have to do without the work input and social contributions of the parent generation. The story of Podinoona and her grandchildren is presented in Box 4. I met them in their home in the outskirts of a Mahaweli village.

**Box 4.**

> Podinoona, a squatter aged 55 years old, had lived here for the last six years. She had run away from her husband who was an alcoholic and used to beat her. She looks old and tired, she wears torn clothes, but the garden is well kept with fruit trees, vegetable plots and nice flowers. Several children are playing in the garden. Originally, she was from a village close to Kurenagala in the south-west, where she used to work as an agricultural labourer and later as a house maid. She has five children, three girls and two boys. The youngest son has passed away, the other one is an electrician and lives close by. Two of the daughters are married and the third one is soon to marry. The elder daughter is in the Middle East (Kuwait). She has been there twice. First, she went as a housemaid and she brought home enough money to pay for the brick foundations for a house. She has gone there this time as a garment factory worker. Her husband is here.
> Podinoona is looking after her five grandsons. Her daughter-in-law, who has three children, is also working in the Middle East (Saudi Arabia), while her youngest daughter works as a housemaid in Dambulla town.
> Podinoona often looks after her son's children as well. The problem is, she feels, that she is too old now to have the responsibility for all these children (maintaining school uniforms, helping with the schoolwork (she is illiterate), preparing food, and tending to the garden,. However, the children help her in the daily chores by sweeping the garden, carrying the water and fuelwood, assisting with the cooking and the shopping, and even sell some of her vegetables in the local market. They only occasionally go to school. It is difficult to make ends meet, and every month they depend on the money sent by her daughter and daughter-in-law. If, for some reason, that money does not come then she and the children have problems and have to borrow money from the local shop owner to survive. Now, they are worried because the money comes irregularly and debts are piling up.
> (Lund, field notes, 2004)

The reality described from the village in the Mahaweli shows how the dynamics of inter-generational relations and child roles are changing amongst the rural poor. It also leads us to revisit how we may understand the global forces at play. What the example indicates is globalisation as *time–space distanciation* (Giddens, 1989; Inda and Rosaldo, 2002; Lie and Lund, 2005). It tells us that the world is shrinking but the same time stretching out. We may experience the world as smaller than before, because it is within reach, directly by fast transport and indirectly via new media. Still, the personal world is at the same time stretching out because social systems are lifted out of the local context, in this case by migration by family members (see also Punch, this volume). This negatively affects the children psychologically, socially and economically. As globalisation works in all directions (North–South, East–West), at all levels (locally and globally), and at a continuously accelerating rate, it directly impacts on people's ability to participate. In the context of the Mahaweli, it can be realised that the global/local nexus is all about (re)creating livelihoods and that the generational support is all about survival and belonging. The agency of the old and the children is determined by the global economy, but it is played out in the village, the local articulation of the global.

## Participation, the Child and the Counter-geographies of Globalisation

I have previously used the concept 'counter-geographies' by Sassen (2000) to show how globalisation may lead to processes of marginalisation of the poor and disadvantaged people in Sri Lanka (Lund, 2000, 2003). Katz (2004), in her recent book on globalisation and children, uses the term 'countertopographies' to indicate the same type of processes. In this perspective, globalisation is seen as a process of differentiation, where individuals, groups or geographical areas are excluded as a result of economic processes of integration. It questions why some actors manage to use the global market to their benefit, while others do not. Processes of marginalisation take place in the wake of globalisation, and profit revenue-making circuits are developed on the backs of the truly disadvantaged. These circuits are what are referred to as the 'counter-geographies' of globalisation, which overlap with some of the major dynamics that compose globalisation.

Empirical evidence from studies on forced migration and trafficking, as well as on child soldiers, supports the thesis of counter-geographies. The thesis also indicates that it is not enough to participate to create social change or develop children's freedoms. When participation is part of a process of marginalisation and exploitation, it becomes irrelevant both as means and as a tool. The example below, of the participation of children as child soldiers in Sri Lanka, illustrates this.

According to Benjamin Hoffman (2003), political violence has two principal forms: direct and structural. Direct violence may be armed conflicts and behaviour intended to do injury to others. Direct violence may also be the result of structural violence. Structural violence is embedded violence: 'It is found in the social, political and economic systems governing the relationships of people. When those structures serve the political objectives of some and leave others to experience their reality as oppressive, that is structural violence. In that way, perceived violence begets violence' (Hoffman, 2003, p. 27).

### *Participation Embedded in Structural Violence: Child Soldiers and the Tamil Tigers in Sri Lanka*

In Sri Lanka, there is an intimate connection between the Tamil diaspora and the Tamil liberation movement. It is widely recognised that the expatriate Tamil community world-wide supports the formation of a separate Tamil state within Sri Lanka and many actively contribute to sustain the protracted conflict in the country. Tamil children are made to participate in this conflict, under extreme pressure and control.

Human Rights Watch (http://www.hrw.org/reports/2004/srilanka1104/) reports that after the ceasefire of February 2002, the Tamil Tiger Movement (LTTE) has continued to recruit children as soldiers (3516 new cases reported in October 2004), including recruitment of children previously released from the LTTE eastern fraction in 2004 (*c*.2000 children), and again after the tsunami in 2005 (number not known). Human Rights Watch and UNICEF have in fact rejected the claim by the LTTE that it has stopped recruiting child soldiers: 'there is still widespread LTTE recruitment of children in the north and east, including the use of threats, coercion and abduction' (*The Sunday Times* 27 February 2005, 7, Box 5). According to the *Daily News in Sri Lanka* (28 February 2005, 11) 'UNICEF reports that between April 2001 to September 2004, there were 4250 reported cases of underage recruitment ... the actual number, needless to say, would be a lot higher'. These standpoints have been proven true: in 2006, during the revival of the conflict, many more children have been abducted (http://www.spur.asn.au/childwar.htm):

**Box 5.**

> Parents who resist the abduction of their children face violent LTTE retribution. This is the story of one girl who was recruited by force in 2003, aged sixteen:
>
> *My parents refused to give me to the LTTE so about fifteen of them came to my house—there were both men and women, in uniforms, with rifles, and guns in holsters ... I was fast asleep when they came to get me at one in the morning ... These people dragged me out of the house. My father shouted at them, saying 'what is going on?', but some of the LTTE soldiers took my father away towards the woods and beat him ... They also pushed my mother onto the ground when she tried to stop them.* (http://www.hrw.org/reports/2004/srilanka1104/ p. 1)

Children are generally abducted while walking between home and school. This has led many parents to keep their children away from school. In other cases, children are abducted during religious festivals or other public events. In fact, parents who could afford it have sent their children to relatives in the southern part of Sri Lanka, or have even sent them abroad to work in the Middle East. Recruitment of children also takes place through sophisticated LTTE propaganda (tales of heroes, parades, speeches, videos, public display of war paraphernalia) or because children have witnessed or experience abuse by the Sri Lankan army (rape of their mothers, interrogations, torture, executions, enforced disappearances) and deprivation and poverty (lack of jobs, education, services). In such circumstances, enlisting in the LTTE has been perceived as a positive alternative to the other options children witness around them.

Once recruited, the children are put in dormitories where they are fed, given uniforms and are trained by senior LTTE officials. They become parts of a regime which allows them to mobilise with other destitute children and they are part of the dominant political force. However, most children are not allowed any contact with their families. The LTTE subjects them to rigorous, often brutal training. They learn to handle weapons, including landmines and bombs, and are taught military tactics. If they make mistakes they are beaten. Children who try to run away are beaten in front of their entire unit to discourage other children from doing the same.

Children, 40% of whom are girls, are recruited to the 'Baby Brigade', and after some time are integrated into other units. An elite 'Leopard Brigade' was formed of children drawn from LTTE-run orphanages and was considered one of the LTTE's fiercest fighting units (UNICEF, 2004). Both girls and boys are used on combats. According to the UN, children were used for 'massed frontal attacks' in major battles, and that children between the ages of 12 and 14 were used to massacre women and children in remote rural villages. Children, often girls, were used in the suicide bombings, because they were less likely to undergo rigorous searches at government checkpoints.

Recruitment of children takes place even though the LTTE has formally agreed upon an Action Plan for Children Affected by War (2003). Transit centres for child soldiers who were returning homes were established by UNICEF. However, according to UNICEF, more than twice as many children are newly recruited as the number who have been released. Hence, the LTTE has failed to meet its commitments to end its recruitment and use of children. Besides, the LTTE practice is a war crime, according to the Convention on the Rights of the Child, to which Sri Lanka is a party. The Worst Form of Child Labour Convention, adopted by the International Labour Organisation in 1999, prohibits the forced recruitment of children under the age of 18 for use in armed conflict as one of the worst forms of child labour.

Thus, child soldiers are not only made to execute direct violence towards others, but their participation in war is embedded in structural violence as well as in fear. In more general terms, participation is about agency and how one performs, but it also about structures and why one performs the way one does. While the Tamil diaspora supports the LTTE, people and communities on the ground are marginalised. They have become victims of geopolitics at the local, national and global levels. The child soldiers in Sri Lanka are being manipulated and abused in the war economy, and they do not participate for the benefit of peace building.

## Deconstructing the Concept of the *Participating Child*

This paper has explored different theoretical discourses in child research and development research. It has been found that there are contingent discourses in development research and child research on the issue of participation, but that these discourses only partly overlap and may actually refer to different processes of social change and development. While child research concentrates on the role of the child to participate as a way to empower self and local communities, recent development discourse focuses on building capacities and strategising to achieve social and economic improvement at different scales.

Furthermore, the discourse on participation has particularly related to praxis and methodology, and is concerned with how people can be mobilised for their own good, on their own or with the help of an external agency that acts as a catalyst in this process. Participation relates to many dimensions and processes. It may be an end and a means, it may be passive or active, inclusive or exclusive, forced or voluntary; it may be an enabling and liberating force and thus empower, or it may be a restrictive force and disempower. Furthermore, participation and empowerment is about sensitising self and others, but it is also about organising the capacity of the individual and a group. The borderline between empowerment and disempowerment is thin and fragile too; what empowers today may not necessarily empower tomorrow. To grasp the complexity of participation, we have to draw on new and critical perspectives on the role of agency in recent development theory as well as insights from post-colonial scholars.

What can be learned from the empirical examples of this presentation is that the representations of children are multiple and diverse. There is a need for child geography sensitive to local, cultural circumstances. When we analyse children's participation to empower, we realise that they are important actors, especially in child-induced activities. We therefore have to acknowledge children's organising powers, as well as their willingness and motivation to mobilise for change. In adult-induced activities, however, much depends on the philosophy and methodologies of the external actor, such as their peers or local institutions.

In rapidly modernising societies such as China, children are perceived as passive recipients of change, and they are deprived of the ability to capacitate. Although children in China seem to fare better than children from impoverished Sri Lankan communities, there is some evidence that Chinese children too are not always looked after all that well. Only in more progressive environments in both countries are children allowed to have more proactive roles and functions.

The two case examples from Sri Lanka show various types of undesirable situations for the Sri Lankan child and their participation reflects different social and cultural positions in Sri Lankan society. In both cases, the significance of children's participation is subject to structural constraints represented by family and local community members, ranging from marginalisation and poverty (the Mahaweli example) to direct violence and fear (the LTTE example). As the participation is embedded in local, cultural and structural constraints, the children cannot effectively empower. Both situations are unwanted.

Participation may also be embedded in external structural forces, such as globalisation. This paper shows that children have become more (not less) vulnerable with globalisation, and participation may result in more exploitation. The reality of children described in the Mahaweli area leads us to conceptualise globalisation as *time−space distanciation*, as the family has 'stretched out' to foreign lands. The family has become dependent on both the home land and the foreign land for their survival. In this process, all family members, including the children, have become important actors in the global economy. So while globalisation has increased the opportunities of some children, it has at the same time restricted the possibilities for so many others.

Another effect of global interlinkages is the Tamil diaspora and the significance it has had in supporting the Tamil liberation movement intellectually and economically. The exploitation of Tamil children coincides with some of the major dynamics that compose globalisation. Globalisation is therefore not only a homogenising force, but a process of differentiation, as some children are integrated in the global economy while others are marginalized, abused or rejected. This is particularly obvious in the case of the child soldiers in Sri Lanka, where participation takes place at the wrong place.

Consequently, since capabilities are not about what children can 'choose' but what they are able to achieve, they depend partly on personal circumstances and partly on social options or constraints. In spite of all that children can achieve, the way they can 'empower' through participation very much depends on social constraints, by parents and/or peers, by the local cultural context and norms, and by the wider, often global, social and economic forces. Thus, a new focus on the 'participating child' implies that the various structural, contextual and geopolitical factors at play will have to be deconstructed to understand the full significance of participation in creating a significant societal and cultural change for children.

### References

Ansell, Nicola (2005) *Children, Youth and Development*, London & New York: Routledge.
Appadurai, Arjun (2001) Grassroots globalization and the research imagination, in Appaduraj, Arjun (ed.) *Globalization*, Durham & London: Duke University Press, 1–21.
Archard, D. (1993) *Children: Rights and Childhood*, London: Routledge.
Booth, David (1985) Marxism and development sociology: Interpreting the impasse. *World Development*, 13(7), 761–87.
CIDA (1992) *How the Other Half Dies*, Canada: CIDA.
Cooke, Bill and Kothari, Uma (2001) *Participation. The New Tyrannym?* London & New York: Zed Press.
Cornwall, A. (2002). Locating citizen participation, *IDS Bulletin*, 33(2), 49–58.
*Daily News*, 28 February 2005, 11.
DAWN (1995) *Markers on the Way: The DAWN Debates on Alternative Development*, DAWN's Platform for the UN Fourth World Conference on Women, Bejing 1995.
Dreze, J. and Sen, A. (1989) *Hunger and Public Action*, Oxford: Clarendon Press.

Escobar, Arturo (1997) The making and unmaking of the Third World through development, in M. Rahnema and V. Bawtree (eds) *The Post-Development Reader*, London & New Jersey: Zed Books.

Friedmann, John (1992) *Empowerment: The Politics of Alternative Development*, Cambridge & Oxford: Blackwell.

Giddens, Anthony (1989) *Sociology*, Oxford: Polity Press.

Hart, Roger (1992) *Children's Participation: From Tokenism to Citizenship*, Florence: UNICEF.

Harvey, David (1989) *The Condition of Postmodernism: An Enquiry into the Origins of Cultural Change*, Oxford: Blackwell.

Hoffman, Benjamin (2003) *Bridge the Knowledge-Action Gap: The Authoritive Statement on How to Reduce Political Violence*, http://newmathforhumanity.com

IFAD (1992) *The State of World Rural Poverty. An Inquiry into Its Causes and Consequences*, New York: New York University Press.

Inda, Jonathan Xavier and Rosaldo, Renato (eds) (2002) *The Anthropology of Globalization. A Reader*, Oxford: Blackwell.

Kabeer, Naila (2003) *Gender Mainstreaming in Poverty Eradication and the Millenium Development Goals*, London: Commonwealth Secretariat.

Katz, Cindi (2004) *Growing up Global. Economic Restructuring and Children's Everyday Lives*, Minneapolis, MN: University of Minnesota Press.

Kjørholt, Anne Trine (2001) 'The participatining Child'—A Vital Pillar in this Century? Nordic Educational Research No 2. Volume 21, Oslo: Universitetsforlaget.

Kjørholt, A.T. og Jens Qvortrup (2000) *Children's Participation in Social and Political Change in western Europe*, Children's Participation in Community Settings, Childwhatch Internation/UNESCO. Oslo, 26.06–28.06.

Lie, Merete and Lund, Ragnhild (2005) From NIDL to Globalization: Studying women workers in an increasingly globalized economy, *Gender, Technology and Development*, 9(1), 7–30.

Long, Norman (2001) *Development Sociology. Actor Perspectives*, London & New York: Routledge.

Long, N. and Long, N. (1992) *Battlefields of Knowledge: The Interlocking of Theory and Practice in Social Research and Development*, London & New York: Routledge.

Lund, Ragnhild (1981) Women and development planning in Sri Lanka. *Geografiska Annaler* Series 63B (1981) 95–108.

Lund, Ragnhild (1983a) The need for monitoring and result evaluation in a development project—Experiences from the Mahaweli Project, *Norwegian Journal of Geography*, 3–4, 169–86.

Lund, Ragnhild (1983b) Working report from the project 'Families' living conditions in the Mahaweli, Sri Lanka', 68 pp.

Lund, Ragnhild (1993a) *Gender and Place. Towards a Geography Sensitive to Gender, Place and Social Change*, Doctoral thesis (Vol 1), Department of Geography, Trondheim. ISBN 974-8202-26-7.

Lund, Ragnhild (1993b) *Gender and Place. Examples from Two Case Studies*, Doctoral thesis (Vol. 2), Department of Geography, Trondheim, ISBN 974-8202-26-7.

Lund, Ragnhild (1994) Development concepts in the light of recent regional political changes and environmental challenges. Doctoral lecture, University of Trondheim, 27 April 1994. *Papers from the Department of Geography, University of Trondheim* No. 139.

Lund, Ragnhild (2000) Geographies of eviction, expulsion and marginalization: Stories and coping capacities of the Veddhas, Sri Lanka, *Norsk Geografisk Tidsskrift–Norwegian Journal of Geography*, 54(3), 102–10.

Lund, Ragnhild (2003) Representations of forced migration in conflicting Ssaces: Displacement of the Veddas in Sri Lanka. *In the Maze of Displacement. Conflict, Migration and Change*, 'Introduction'. Høgskoleforlaget, pp. 76–104.

Lund, Ragnhild and Agyei-Mensah, Samuel (2007), Queens as mothers: The role of traditional safety net of care and support for HIV/AIDS orphans and vulnerable children in Ghana, *Geografiska Annales, Series B, Human Geography* Issue 3.

Mehmet, Ozay (1995) *Westernizing the Third World. The Eurocentricity of Economic Development Theories*, London & New York: Routledge.

Mohanty, R. *et al.* (2006) *Participatory Citizenship: Identity, Exclusion, Inclusion*, New Delhi: Sage.

Moono, M.M. (2006) *A Study of How the Lusaka SOS Children's Home Rebuilds the Orphans and Vulnerable Chilren's Lives and Capabilities*, MPhil thesis in Development Studies, Department of Geography, NTNU, Trondheim.

Mueller, H.P. and Hettige, S.T. (1995) *The Blurring of a Vision—The Mahaweli*, Colombo: Sarvodaya Book Publishing Services.

Mwale, Masauso (2001) *Vulnerable Children and Non-Governmental Organisations. An Exploratory Study into Urban and Rural-Based NGO Intervention Methods and Their Impact on Children*, MPhil thesis, Department of Geography, NTNU, Trondheim.

Mwene, C. (2006) *An Assessment of Community Participation and Empowerment through Non-Governmental Organisations' Development Work among the Rural Poor*, MPhil thesis in Development Studies, Department of Geography, NTNU, Trondheim.

Nerfin, Mac (1977) *Another Development—Approaches and Strategies*, Uppsala: The Dag Hammarskjold Foundation.

Nussbaum, Martha (2000) *Women and Human Development. The Capabilities Approach*, London: Cambridge University Press.

Pieterse, Jan Nederveen (2001) *Development Theory. Deconstructions/Reconstructions*, London: Sage.

Punch, Samantha (2007) Negotiating migrant identities: Young people in Bolivia and Argentina, *Children's Geographies*, 5(1–2), 95–112.

Qvortrup, Jens (1997) *Constructing and Reconstructing Childhood: Contemporary Issues in the Sociological Study of Childhood*, London: Falmer Press.

Robertson, Roland (1992) *Globalization: Social Theory and Global Culture*, London: Sage.

Roche, J. (1999) Children: rights, participation and citizenship, *Childhood*, 6, 475–93.

Sassen, Saskia (2000) Women's burden: Counter-geographies of globalization and the feminization of survival, *Journal of International Affairs*, 53(2), 503–25.

Sen, Amartya (1981) *Poverty and Famines: An Essay on Entitlement and Deprivation*, Oxford: Clarendon Press.

Skelton, Tracey (2007) Children, Young People, UNICEF and Participation, *Children's Geographies*, 5(1–2), 165–181.

Stephens, S. (1994) Children and environment: Local worlds and global connections, *Childhood*, 2, 1–21.

*Sunday Times*. 27 February 2005.

Tuwai Smith (1999) *Decolonizing Methodologies. Research on Indigenous Peoples*, London: Zed Books.

UNICEF (2004) *The Invisible China: The Situation of Women and Children*, Bejing.

United Nations Development Programme (2004) *Human Development Report 2004*.

Van Beer (1995) *Participation of Children in Programming*, Stockholm: Swedish Save the Children.

Waters, Malcolm (1995) *Globalization*, London: Routledge.

Williams, Emma (2004) *Children's Participation and Policy Change in South Asia*, CHIP Report No. 6: London.

Winsvold, Aina (2004) *Når arbeidende barn mobiliserer seg. En studie av tre unioner i Karnataka, India*. Lund Dissertations in Sociology 62, University of Lund.

## Internet sources

http://www.hrw.org/reports/2004/srilanka1104/
http://www.spur.asn.au/childwar.htm

# Embedding the Global Womb: Global Child Labour and the New Policy Agenda

OLGA NIEUWENHUYS

*Department of Human Geography, Planning and International Development Studies, University of Amsterdam*

## Introduction

This paper seeks to understand how representations of child labour shape children's lifeworlds globally. I use the term *lifeworld* to qualify the social world, which unevenly distributed symbolic resources contribute to misrecognize in favour of a mythical global order (cf. Couldry, 2003, 41 ff). My suggestion is that as representations of child labour hold out the promise of a labour-free childhood to the worlds' children, children's day-to-day responsibilities and work routines are diminished, denied or even criminalized. These routines are typically located in hidden landscapes of reproduction in the global south.

The wider issue informing my argument is the 1990s insistent appearance in international development policy of references to child labour. I suggest that the timings are not by chance. The unearthing of this typically nineteenth century northern policy issue signalled that something fundamental had changed. But rather than in the everyday practices of children in the developing world the change had been, I believe, in the northern project for the south. My argument seeks to retrace the temporal and spatial itinerary of what, to distinguish it from its earlier historical form, I term global child labour (see also Rahikainen, 2001) and runs as follows: With the fall of the Berlin Wall in 1989 and the ensuing triumph of neo-liberalism, a global restructuring of production and reproduction superseded the post-war development project. In the developing world, global restructuring upset earlier post-colonial childhood policies, dramatically changing the policy environment for children. In the new policy agenda there was to be only one, global childhood, the preserve of a minority of children symbolically participating in the market as autonomous, liberated consumers. The image of the non-consumer child, partly born of draconian cuts in social expenditure imposed on the south and partly ideologically constructed, came to symbolize two things: on the one hand it added splendour to the staggering levels of consumption that globalisation ushered in for a minority of privileged children, on the other it justified intervention in the lives of the excluded majority in the global south. Children whose lifestyles contradicted the global ideal were represented as lacking in something essential. Little mattered how much their lifeworlds articulated enduring social relationships between the generations and how much had been done to rebuild these relationships on the ruins of the colonial world. Representations of global childhood revolved on the 'best interest of the child', defined as an individualized self-interest to become a full participant in the new market-approach to development. Here I propose to see global child labour as part of the rituals set in place to stake out the new field of consumer childhood and to explore what these rituals have entailed for the non-consumer child, particularly in terms of the symbolic violence exercised to discipline children's lifeworlds.

I start the paper rethinking historical child labour and suggesting that its abolition in the course of the nineteenth century should be set against the paradox of territorialized nation-states emerging in a context of extra-territorial empire building. Abolition crucially misrecognized the lifeworlds of children in the occupied colonial territories. Secondly, I probe into the discontinuities between historical child labour in the industrial north and today's global child labour by shortly discussing how development theory engaged with the issue of childhood in the post-colonial world. Reinventing childhood as the inevitable outcome of enlightened free-market economics, I thirdly claim, negotiates northern anxiety about an unfolding race to the bottom with the interests of southern elites in maintaining their comparative advantage. This is particularly evident in the ways child labour rituals are framed to justify southern elite's opting out of public policy to ensure the reproduction of the vast reservoirs of cheap labour in their territories. The tacit policy understanding is that if draconian cuts in social spending generate child exploitation, the responsibility would entirely lay with the community's backward culture and lack of understanding of the true nature of childhood. I finally argue that in the new social order children's lifeworlds are trapped in the logic of a self-reproducing workforce for which rituals of global child labour abolition summon up the ever-receding mirage of a better life.

## Rethinking Historical Child Labour

In a sense, there was nothing new in the way international development agencies came to restage child labour in the 1990s. Global child labour would be eliminated with means that had already proven their worth in nineteenth and early twentieth century industrial societies, so was the official belief (Weiner, 1991). My point here is that this belief ignores the existence of vast territorial dominions lying outside the borders of nation-states where the issue of child labour was simply not allowed to arise. Broadly re-examining from a global perspective the elimination of child labour in industrial societies, I suggest here that elimination may be interpreted as a way to deal with what 'development' had, paradoxically, made problematic: the production of life. I first dwell shortly on the neglected role of reproduction in industrial societies and maintain that the mainstay of child labour policies was to secure children's free work in and around the home and in childcare. Secondly, I suggest that as these policies turned out to be relatively costly in the industrial areas, with their free work children in the colonies subsidized abolition not only of child labour but of large families as well. Childhood became, thirdly, the ideological justification for territorial distinctions between areas in which children would have been saved from child labour and areas where they were subjected to new regimes of work that denied the likelihood of their being exploited.

Rahikainen forcefully argues that child labour eradication in industrializing nation-states should be reconsidered in the light of the wider context of international markets in which they operated (Rahikainen, 2001). She suggests that as opportunities to push child exploitation to the margins of the unfurling world economy grew, child labour could be abolished in the core areas. This does however not explain why the removal of children from employment had become an issue sufficiently important to inspire a vast child saving movement (Cunningham, 1991). I suggest that the missing link is the place of reproduction in industrial society. Dispossessed peasants no longer needed their children as living repositories of family wealth, for children could not repay their parents back by working on the land they no longer held. Industrial wages being barely sufficient to maintain a single worker, the elderly, the sick, the unemployed, but above all children, turned into a 'social problem'. Though children were put to work as soon as possible, they seemed seldom able to earn enough for their maintenance (Rahikainen, 2004). The Lancashire cotton mills at the end of the eighteenth century that employed tens of thousands of children had soon discovered the disadvantages of child employment. Children were freely available from orphanages and poor houses, but employers had to feed, cloth and house them and this proved a costly affair. Children also formed an unstable labour force running away if offered better opportunities elsewhere. Sons and daughters of small peasants would only be available during the off-season and would return to their father's homestead if needed there. They were also not as productive as often believed, the only way for the employer to recover the expenses of their maintenance being labour-tying contracts until the age of 21 (Lavallette, 1994, pp. 190–3; Rahikainen, 2001). Mill owners were gradually to discover that it was more profitable to employ adult workers, for as long as adults could kept a foothold in subsidence farming children could negotiate relatively good wages. The situation drastically changed by the end of the nineteenth, when men took the place of children and both women and children could be excluded from industrial employment. Men's wages stabilized at the level of a 'family wage' in which allowances were made for the subsistence of women and a relatively small number of children, guaranteeing the net reproduction of the working class family. The payment of the family wage was crucially premised on the massive availability of cheap colonial commodities (tea, wheat, cotton, wool, etc.) that significantly enhanced its buying power but as well on the displacement of labour intensive food and raw materials production to the colonies (Wolf, 1982).

This brings me to the fate of children in the colonies. History is singularly silent on this issue. Their perceived agrarian nature and subsidiary economic role would have ruled out the likelihood that child labour existed there. My suggestion is that, by the times colonies were turned into providers of raw materials and cheap goods for the workers' family, the idea had firmly been established that regulating reproduction was crucial for the stability and profitability of industrial production. Great pains were taken to preserve or, if needed, to reinvent, the peasant family as the locus of self-sufficient reproduction (cf. for an example Mamdani, 1996). A modest body of postcolonial research evinces that in the colonies children worked hard from a tender age and that their work articulated ubiquitous exploitative patterns between colonizer and colonized. Far from ever being effectively addressed, this work was simply not allowed to emerge as a policy issue. Tellingly, even as late as 1919, the first ILO Convention on child labour conveniently excluded from its definition virtually all forms of colonial child work: 'child labour' did not include work outside the industrial sector, forced labour, work that was unpaid or carried out within the ambit of the family business, the peasant holding or the household. Both justification for the civilizing mission of the mother country and the smooth functioning of colonial administration required that overt competition over labour, even among the colonizers, be either kept strictly in check or be hidden under the veil of tradition, socialisation, education or corrective disciplining.

To disguise child labour on their plantations, planters-cum-missionaries in Swaziland, for example, changed the identity of their child labourers into school children by offering some form of free education in the off hours of work (Simelane, 1998); migrant child workers were redefined as 'stray children' in former Rhodesia and were assigned as servants to a white master supposed to teach them Christian values (Chirwa, 1992; Greer, 1994); in cities of today's Bangladesh migrant working children were similarly labelled 'street children' and interned in workhouses (Balagopalan, 2002); to evade prohibition on the work of the thousands children slaving day and night in the tobacco factories the Netherlands Indies (today's Indonesia) ILO definitions of factory, day-time and even 'child' were tropicalised. In the tropical climate, the day would be too hot for work, so that children would find no harm in working longer after sunset; similarly the 'oriental' child would mature faster, so that minimum ages could safely be lowered (White, 2001); South Indian children working from the age of three or four with their mothers in the manufacture of coir yarn to feed European mills were similarly set as example of the advantages of *preventing* child labour. The underlying justification was that the Oriental family needed household activities to keep rural women and children occupied. As an early example of the theory of 'competitive advantage', the coir yarn sector would demonstrate how child labour prevention enabled primitive forms of manufacture to survive in the face capitalist competition, guaranteeing hereby both social stability and plenty of docile, cheap labour power (Nieuwenhuys, 1994).

What this admittedly patchy evidence suggests is that child labour elimination in Europe and the USA went hand in hand with new forms of child exploitation unfolding in the colonial dominions. In other words, the success of the transformation of childhood into a perceived period of innocent play and study, owned much to its territorialisation. Unfurling childhood as the primordial site of national identity hinged not only on child labour abolition, but universal, compulsory schooling inculcating a common language and culture to the 'becoming' generations (Gellner, 1983; see also Libal, 2002). Childhood being territorially circumscribed to the colonizer's home countries, in the colonial world children and adults merged into the indistinct category of the 'Other', people whose cultures and traditions made them fundamentally unknowable. The national borders of pre-WWII Europe can therefore arguably be read as symbolic markers of child labour-free zones justifying colonial projects in which very young people could be exploited at will because, in a way, they were *not really* children. The 'Other' was represented as growing up in lifeworlds situated in mythical traditional villages where daily life would be steeped in culturally meaningful rituals and subsistence practices (see Chirwa, 1993; Greer, 1994).

As after WWII newly independent nations born from decolonization started upon a course to catch up with their former colonial rulers, the latter had to reinvent a legitimate role for themselves. This they found in promising a brighter future for the coming generations of their former colonies though 'development'. At the end of the path to 'development' shone, as I now turn to contend, the promise of a multiplication of national childhoods in a postcolonial world.

## Paradox of Development

The gap between the end of historical child labour in the north around 1930s and the emergence of its global form in the 1990s, spans nearly 60 years. To understand why child labour re-emerged, a short incursion in the history of 'development' is illuminating. The question here is this: How to explain the paradox that 'development', that started holding out a brighter future to the post-colonial generations ended up producing fifty years later images of sweated children reminiscent of Britain's industrial revolution? My contention is that global child labour reiterates the fundamental difficulty of 'development' to resolve the issue of reproduction and signals that major shifts towards new forms of labour control are underway. Here I do little else than roughly discern three phases in the history of development: (a) decolonisation (1945–1970), (b) transition years (1970–1990), and (c) globalisation (1990–today). My aim is to seek to understand why global child labour paradoxically signalled the end of development as an attempt to establish territorialized national childhoods in the former colonies.

After WOII the USA unfolded a vast political and economic programme of intervention in the former European colonies and protectorates. Introducing ideas of equality, liberty and rights, US interventions differed substantially from the old colonial paradigm based on perceived differences in levels of civilization and race between the colonizer and the colonized. These ideas were expressed in the *Declaration of Universal Human Rights* that accompanied the founding of the UN in 1948 (see also Rist, 2001). The Declaration differed however from the European and American constitutions on which it was framed in that, rather than protect citizens from interference from the state, it prescribed the United Nations' duty to intervene in the lives of citizens in the developing world to provide them with their economic and social rights (food, education, health care, welfare, etc.) In other words, the Declaration laid the legal foundation for intervention in the newly independent nations in the name of development (see also Koshy, 1999).

In the first two decades of development, and in spite of extremely modest outlays, the idea that the north had to help the south develop gained growing ascendancy. A plethora of messy issues—racism, slavery, illiteracy, caste discrimination, landlessness, malnutrition, exploitation, child labour—which development promised to solve meshed in its antipode, poverty. Underneath the cloak of poverty, however, was hidden a fundamental presupposition: poverty would be nothing but the lack of a very specific sort of capitalist wealth. Forms of wealth departing from this form—such as natural resources, local institutions, knowledge passed down from generation to generation, reciprocal exchange of goods and services and so on, were treated as inconsequential if not as barriers to development. As development promised to develop a market of skilled labour, children's roles in the reproduction of non-capitalist wealth continued, as under colonial rule, to be treated as a sign of backwardness or 'resistance to change'. With wealth reduced to what counted in the north, the south was presented with no other option than to seek to follow the pattern of development set out for it. Without this vision of development, the term poverty was indeed meaningless (Escobar, 1995; Rist, 2001).

The 1970s oil crisis put an abrupt end to import-substitution strategies meant to make the new nations less dependent on export of raw materials and simple manufactures, the two sectors in which most of the population—including children—worked. Their reliance on these exports would have been the cause for the secular deterioration of their terms of exchange with the north. The Bretton Woods institutions (IMF, World Bank, etc.), founded in 1973, offered to recycle the mass of petrodollars that became available from the oil producing countries into loans to developing nations' governments struggling to maintain their energy supply. This gave the institutions a vantage point in forcing the soon heavily indebted southern governments to reform their economies in the interest of northern capital. In virtually all countries of the south these so-called *Structural Adjustment Programmes* (SAPs) entailed dramatic cuts in social spending and the roll back of the state. Protected food prices, feeding programmes and free primary education and health for the mass of children became gradually a thing of the past (Bradshaw, 1993).

Within two decades of the oil crisis most countries of the South had entirely abandoned earlier programmes of import-substitution and had set the door open to northern investments. The social sector had as much as possible been privatised, funds from the north being increasingly channelled through NGOs. As the market was left to perform the promised miracle of economic growth, social programmes were reduced to 'targeting' the most vulnerable. I return to this point in the next section.

What concerns me here is that at this juncture the global child labour issue starts appearing on the developmental stage. It is not the place here to go into the details of global child labour policy. Suffice to say that the issue gained international prominence during the founding ministerial meeting of the WTO in 1996 and soon received enthusiastic backing from Northern governments, exporters, trade unions and consumers. To channel the enthusiasm, the ILO was revived and received such substantial Northern funding for its child labour programme IPEC (*International Programme for the Elimination of Child Labour*) that, by the end of the twentieth century, global anti-child labour campaigning overshadowed other activities of the organisation (see also Myrstad, 1999; ICFTU, 2000; Fyfe, 2001; Scheuerman, 2001; Burgoon, 2004). In this campaigning, global child labour was overwhelmingly interpreted as a repetition of northern history. This interpretation is misleading insofar that historical and global child labour differed markedly on several accounts: First, historical child labour was premised on a strict cognitive separation between national childhood and the Other, unknowable childhood. As national borders set the physical limits to the outcry against children's exploitation, the issue remained an inward looking exercise. The role of the Other was to justify exclusion. Global child labour policies are radically different in their construction of the Other. They have originated from northern-dominated institutions and agencies as a diagnostic discourse about a social gangrene affecting newly independent nation-states in the south. As northern child labour seated uncomfortably with the issue it was carefully kept aside, national borders marking the limits of what was a deeply outward-looking approach. Second, public awareness about the evils of child labour in the north had articulated support for social policies. With global child labour the diagnosis coincided paradoxically with the imposition of an economic regime upon the south that effectively dismantled social policies and in particular scaffolding childhood institutions such as food security, free and universal education and free basic health. Third, though the discourse was cloaked in a language that suggested that a nineteenth century northern ghost had stood up from its grave, the reality was dramatically different. Historical differences between the industrialized world and what had become the developing world in the nature and incidence of child labour largely persisted. With a child population of roughly 400 million in 1980, India had for instance records of only 15 million children in employment, while the occupation of at least a 10-fold working in agriculture and in the informal urban sector was unaccounted for its failure to qualify as 'child labour'. So did of course unpaid help, domestic work and childcare. In other words, with global child labour northern public opinion gathered against a monster that was largely of its own making. The exploited child in the developing world became the spectre unearthed from its historical grave to haunt the north's economic security and welfare system. As it took the familiar form of an issue that would have been dealt with successfully in the past, it was a reassuring reincarnation ritually framing the new global policy agenda for the post-development south.

To now sum up this section, child labour was absent from the development agenda as long as postcolonial nation-states were either building their own national childhoods or managed to maintain the illusion that they were doing so. This changed when, with the triumph of liberalism the post-war development project effectively came to an end and trade barriers started giving way to a global market. It was then no coincidence that the child labour issue re-emerged simultaneously with the liberalisation of the world economy. First, it had existed before, but was ritually constructed to hide away the new forms of child exploitation unfolding in the occupied territories that formed the colonial world. A new geo-political formation, the colonial state, effectively justified this exclusion. Second, when it re-emerged, the international policy agenda was firmly set against post-colonial nations' attempts at protecting their populations with trade and investment barriers. International financial institutions left no doubt that their teeming populations working for the proverbial bowl of rice represented the competitive advantage that would permit them to survive the transition to globalisation. And thirdly, threats to northern welfare in the form of mass migration and de-localisation needed to be directed towards a recognizable enemy. Collective memory about the role of child labour eradication in the making of the welfare state was sufficiently alive to align political response behind a common child saving agenda. To clarify why this agenda won legitimacy not only in the north but in the south as well, I now to turn to the power elites for whom global child labour offered distinct advantages that the post-colonial paradigm failed to do.

## From National to Global Childhood

Global child labour policies were not merely imposed on the south. As elites in the post-colonial nations, voluntarily or under threat, gave up their ambitious project of catching up with the north, they pragmatically chose to compromise to remain in the game. Under enormous pressure from the Bretton Woods institution to eliminate protected food prices and make inordinate cuts in childhood-related spending, they needed urgently to find ways to justify these highly unpopular measures. This may explain, I believe, the unprecedented urge with which even governments notoriously failing their children ritually underscored that they were jumping on the liberalization train by ratifying the 1989 *UN Convention on the Rights of the Child* (Nieuwenhuys, 2001). My contention here is that ratifying the CRC signalled that the old national childhood paradigm had been abandoned in favour of what Pupavac pointedly terms a children's rights regime (Pupavac, 2001). Understanding this regime is critical to explain why the exposure of global child labour failed to question draconian cuts in social spending. Even if southern governments certainly protested vigorously against the cuts, in global campaigning the link between these cuts and child labour was seldom made. The slogan of the day was rather TINA: There Is No Alternative. Authoritative policies coming from the IMF, the World Bank and, from the mid-1990s, also the WTO and the ILO, paid at most lip-service to measures such as debt relief, protection of southern markets or global taxing to support developing state's need to protect children against the effects of liberalisation on their economies. The tacit mood was that it was simply unavoidable that children would pay with their lives what were evasively termed 'economic reforms', 'exogenous shocks', 'vulnerability', 'risk-aversion' and so on (Cornia, 1987; World Bank, 2001). Problems were attributed to the earlier inordinate ambitions and corruption of nationalist, socialist or communist post-colonial elites. Given the laws of the market, consensus about the state's role in child labour converged around another set of practices: first, at the level of ideology, ritualistic celebrations of children's rights, second, disengagement with social justice and, finally, financial support to projects and programmes aimed at reforming the children themselves under close surveillance of northern donors.

When global child labour policies unfurled, they did so through the language of rights at the ideological level and at the practical level through NGO projects and programmes carried out in close cooperation with northern donors. The new policy agenda was already well in place. Child labour was no longer to be the concern of the state alone: but of parents, communities and global civil society as well as the children themselves. The agenda to combat child labour did therefore not involve much more than applying what was already foreseen in the CRC: advocacy, training and lobbying in addition to detecting criminal cases and mobilizing the media and the judiciary. As trade unions failed to make core labour rights conditional for WTO membership, northern governments directed financial funds at the ILO in view of drafting a new child labour convention that would specifically address the problems arising from globalisation. The result was the 1999 ILO Convention 182 on the Worst Forms of Child Labour, a document that curiously conflates crime with work and, though defended as a necessary addition to earlier child labour Conventions, in practice may in the long run replace them.[1]

The UN Convention on the Rights of the Child (CRC) that was virtually unanimously ratified from 1989 onwards, met with surprisingly little dissent. The CRC offers, as many have argued, a contradictory blend of entitlements, rights and wishful thinking that is the typical product of UN compromise. The overarching idea is that as children's rights are put into practice they will be able to both accommodate cultural diversity and help local cultures evolve towards universal notions about the best interests of the child. As it offers but a broad framework, not a set of hard laws to be enforced without discussion, the CRC provisions have unsurprisingly been interpreted in the light of prevailing political economy in which the state is reduced to the role of guarantor of a level playing field and the onus of child protection is placed on the shoulders of parents. The quality of their children's childhood is now envisioned as depending upon their ability to fully participate in the market.

A special committee reviews periodically developing countries' progress in implementing the provisions of the CRC. Notoriously failing governments may be verbally heavily criticized, but are not subject to other sanctions. What is more serious, even if Bretton Woods institution and WTO policies, or for that matter major donor countries in the North play a growing role in shaping policies in the global South, they escape reviewing. Nor are MNCs or private banks submitted to scrutiny. Debt servicing, the retreat of the state from social spending, the creation of a liberal climate to attract investors and investments in infrastructure and security are all treated as constraints beyond discussion within which children's rights will have to be accommodated. Within these constraints the room for manoeuvre is of course very small and is essentially a matter of good intentions. Implementation on the ground has been mostly left to human rights or children's rights NGOs who are likely to be more accountable to northern donor agencies than to governments or the people they say to represent. The overarching idea is that NGOs are best equipped to mobilize means and people from a variety of sources—mostly private— through training, lobbying, and advocacy. For those seriously in danger, or perhaps more precisely those who pose a danger, special targeting programmes have been set in place, preferably through the voluntary sector. Targeting has the triple role of rolling back the state, building barriers against the demands of the poor for social goods and services and opening up fields for philanthropic entrepreneurship. Tellingly, the vast majority of children targeted by NGOs are teenage urban males. The very young, girls and those living in distant rural areas are generally unable to access their interventions.

The 'model' NGO-child is preferably sufficiently small in size to appear much younger, has a heart-rending personal story of extreme abuse that refracts negatively on both his or her own society and parents, has the determination and intelligence of an autonomous adult yet nourishes only one dream: be restored to a lost childhood. The persistence of this model among the clients of NGOs across the south intimates that child rights projects may fruitfully be conceived as performances where carefully selected child-actors play out scripts ritually foregrounding the spectacle of global childhood. Local conceptions, culture and practices are treated as the Oriental Other, the 'before' on which the child actors are invited to dwell *in extenso* to contrast its hellish inhumanity with the 'after', where children attain both their rights and their salvation (cf. also Hecht, 1998). The script feeds on a common policy language which NGO-personnel can learn, understand and practice through a multitude of international conferences, training sessions and policy meetings in mutual supportive exercises that help stake out the global childhood policy field. Training packs, project design formats, evaluation protocols and financial accounting procedures further help create a truly globally shared culture about how to deal with children's issues. The spectacle, finally, is routinely performed for visiting (potential) donors, in the media and during events that celebrate the great strides made in the advancement of children's rights. Even, or, as I personally witnessed during a 'street children ballet' at the national theatre of Addis Ababa in the late 1990s, perhaps especially, when the government is under great pressure to cut down on the last vestiges of what could make the lives of children minimally bearable.

There have been distinct financial advantages for southern elites to jump on the children's rights train once it set in motion. Busying the language of rights has become the key to access important funds being channelled through donor NGOs. These are typically represented as ideologically independent, non-political, non-confessional, non-governmental organisations that are financially unable to survive international competition without sizeable northern state support (Tvedt, 2002). Major children's NGOs in The Netherlands (for ex. *Plan International*, *Terre des Hommes* and the *Foundation Children's Stamps*) are, for example, government funded up to 70%. The same is the case with the Norwegian and the Swedish branches of *Save the Children*. The US, for their part, have traditionally channelled most of their aid through NGOs (CARE being no doubt the classical example), and under the Bush administration have increasingly done so, particularly when it came to the evangelical ones (Hearn, 2002).

The privatisation of international aid turned out in the interest of ruling elites not only because NGOs paid significantly higher salaries and offered better career opportunities than the public sector but because they opened up to the emerging world of global childhood expertise on the one hand, and on the other helped feed clientelistic relationships with followers guaranteeing to elites a solid footing in local politics (Hecht, 1998; Michener, 1998; White, 1999; Ulvila, 2002). For heavily indebted southern government elites staying in power now increasingly means taking advantage of the new opportunities and finding a niche as intermediaries between the local poor and northern donors. Political legitimacy is no longer gained through some encompassing but difficult if not impossible policy of distributive justice, but through the forging of dependency ties, fed by global financial networks, with needy people.

Global child labour policies met both northern organised labour interests and those of southern elites: for organised labour the main point was to eliminate direct competition that would take away jobs. Reinventing the colonial division of labour between the industrial north and the agrarian south was hence the mainstay of policies. This becomes apparent, as I shall develop more in depth in the next section, from the justification sought for ignoring or dismissing children's work in the domestic arena, in the non-criminal informal sector and in agriculture.

For the southern elites the stakes have been high. Accepting that there would be no childhood without a global market and global childhood being now patterned on the northern consumer child meant essentially that austerity and justice no longer went hand in hand: to realize their child rights individual children can now legitimately consume ostentatiously even in the midst of acute deprivation. Nowhere does the CRC indeed state that this right would be predicated upon a principle of solidarity with less fortunate children. From now on enterprising parents earning money to spend on their children's good quality food, education, health, housing and recreation, in other words, on the realization of their children's rights, were to be set as examples. The failure to realize rights became a 'failure of parental altruism' and put on the account of poor parents' lack of initiative, ignorance, laziness, insensibility or, worse, religious and political fanaticism. A failure to offer constant care and supervision, to buy private education, to provide children with expensive medicines, vitamin-rich food, electronic toys and mobile phones and so on, was no longer the outcome of some structural injustice in the global system of distribution. Ruling elites could unfurl the familiar interpretation of poverty as resulting from the vicious circle of high fertility, early labour and ignorance (see for India, Ramanathan, 2000), and offer no more than support to empower people to 'help themselves'.

A rights approach that shifted the focus from the state as distributor of public goods to the child as endowed with abstract universal rights, disembedded from the wider context, helped justify the state's disengagement with social and cultural reproduction. As I will now turn to discuss, this disengagement was not only in the direct interest of the southern elites who acquired power as mediators of northern funds. It was more generally conducive to guarantee the stability of the 'competitive advantage' of the global south at the heart of the new international political economy.

## Disciplining the Global Womb

Its vast reservoir of cheap labour is arguably the essence of the 'competitive advantage' that enables Southern elites to participate in the globalised economy (Fallon, 1998). In this section I argue that this advantage is, as of old, contingent upon preventing children from being directly employed in the labour market.

Though in-depth studies have remained scarce, anthropologists have recurrently analysed children's work in the context of local cycles of exchange. In the early 1980s, Pamela Reynolds worked for instance among the Tonga of the Zambezi valley in Zimbabwe, a starving population whose economy seriously suffered from the combined effect of a lack of water (due to the construction of a large dam) and the exclusion from the use of commons for hunting (due to the creation of a wildlife sanctuary) (Reynolds, 1991, p. xxiv). As their work is intertwined with adults' and with the local structure of subsistence farming, Reynolds contends that children:

> ... provide the adjustable labour during periods of intense farming activity, and that women depend upon their children's assistance. Women and children perform as work units and it is these work units that are sometime called upon by men to provide labour in the field. Women direct children's labour. While many children do not work as hard as adults nor achieve as much, they perform other duties at the fields such as the preparation of meals and guarding of crops that, were it accounted for, might balance the labour accounts. Children under ten are kept busy attending to infants and toddlers. (Reynolds, 1991, p. 53)

My own research on girls' 'invisible' work in the production of coir yarn in a Kerala coastal village corroborates that this work is often part of a production process that relies heavily on the need of poor women to feed their families. When men are thrown out of work or migrate to distant fishing grounds the family relies mainly upon the earnings from coir yarn to make ends meet (Nieuwenhuys, 1994, 121 ff). Few girls are aware that they actually contribute, as I was able to compute during my field-work, between 60 and 70% of the labour necessary to make the finished product. In spite of this, even the poorest household's yearly income from coir yarn rarely exceeds 10% of their total earnings. The surprising survival of this ancient handicraft is only possible, I argue, because of the availability of girls working for inconsequential returns that are acceptable because they are enhanced by the moral value enshrined in the work. But this labour of love in no way stands in the way of the coir business' prof-itability, and this is crucial. For a holder of capital, investing in coconut husks is just as profitable as any other investment in the rural economy, and enables to realize profits as high as 10–15% a month. These profits are only possible because of poor women's and girls' heavy dependency on the manufacture of coir yarn for a living, which compels them, in the absence of alternatives, to buy husks at prices that severely curtail the income they can realize after they have been processed into yarn. Not that husks are rare or highly valuable: in those parts of India where there is no coir making, husks are used as cheap fuel, and near to worthless. It is then the labour of love that they are capable of absorbing by being turned into coir yarn by poor women and their daughters that turn husks into alluring objects of investment and enables the local dealers to realize their profits. The economy of the village can clearly not cope without the insertion, on other than purely economic grounds, of girls' labour of love in the making of the coir yarn. This insertion is itself part of a larger family strategy that supports and favours the successful competition of antediluvian forms of pro-duction against the threat posed by the rationality of the market.

Recent anthropological studies confirm this pattern of intergenerational sharing of responsibilities and work allocation in rural societies in Guatemala (De Suremain, 2000), Mexico (Bey, 2003; Taracena, 2003), Togo (Lange, 2000), Ghana (Van Haer, 1982), Nigeria (Schildkrout, 2002; Robson, 2004) and South Africa (Levine, 1999) as well as urban communities in for example northeast Brazil (Hecht, 1998; Kenny, 1999), Thailand (Montgomery, 2002), Peru (Invernizzi, 2001) and India (Swaminathan, 1998; Bukhut, 2005). These studies substantiate that children's work cannot be under-stood in isolation from the totality of activities that make up local economies and that it appears far more crucial in the lives of the poor than common wisdom holds it to be. These studies also substantiate that employment is not the ubiquitous way the children of the poor are exploited. Children remain firmly attached to their primary role as carers, while also providing very essential services in the sphere of subsistence and helping produce the raw materials and manufactures that drive global economic growth. It is then precisely its being couched in the moral economy of the family, with its preoccupation with subsistence and the preservation of life rather than with economic gain, that makes for the ubiquitous way in which children's work is embedded and acquires its meaning.

What are we then to make of the avalanche of casework bringing the global sweatshops, where nimble fingers would provide the ultimate source of capitalist gain, to the limelight? The problem is that this casework tells us very little about the incidence of sweated child labour while it is so overshadowed by moral outrage to become virtually useless to the more serious researcher. This type of casework is chiefly concerned with walling off acceptable from unacceptable forms of child work and tells us very little about the life-worlds in which they are embedded. As if children's everyday toil would be too banal to warrant attention, the spotlight is on the exceptional, extreme cases. Why this is so is, I feel, chiefly a matter of establishing an illusionary social order that naturalizes chil-dren's lifeworlds in the global south. Striking largely unconscious emotional chords, global child labour touches on sensitivities belonging to the collective imagination of the groups they address. The collective memory about child labour in the north is alive and its message is straightforward: Global child labour is both source and symptom of untold misery. Economic growth, welfare and consumption would only be possible if children are removed from the labour market. The ILO eloquently translates the emotions in the following way:

> Child labour is clearly detrimental to individual children, preventing them from enjoying their childhood, hampering their development and sometimes causing life-long physical or psychological damage; it is also detrimental to families, to communities and to society as a whole. As both a result and cause of poverty, child labour perpetuates disadvantage and social exclusion. It undermines national development by keeping children out of school, preventing them from gaining the education and skills that would enable them as adults to contribute to economic growth and prosperity. (ILO, 2002, p. 1)

What is proposed here is in fact a simple recipe based on the idea that the south can copy what the north achieved. Enough has been written on how a growing inequality in the distribution of worlds' resources feeds the orgy of consumption among the happy few both in the north and among the elites in the south while exacerbating poverty among children for me to repeat it here. What is more important is to read today's globalisation process and its preoccupation with child labour as a renewed engagement with the problematic of life production under capitalist conditions which emerged in the industrial north during *The Great Transformation* (Polanyi, 2001). Arguably, if children in the global south would massively engage in paid labour on any significant scale, adults would have to be paid more, as they would need to buy food, piped water and cooking gas, and, what is more important, hire nannies and nurses in addition to save for their old age and pay for their children's education. In the ethnographic studies mentioned above, there is sufficient evidence to suggest that in many part of the rural south children contribute to the maintenance of teachers and school buildings by offering free labour on school gardens or during school improvement days. Were this work unavailable the costs of primary schooling would dramatically rise. Because children will, in time, also offer old-age support and produce new children for care-work of the elderly, the whole issue of pensions does not even arise. Were children only to work against payment, raising them would, in short, undermine the competitive advantage that drives today's globalization.

Globalisation has opened up ways of tapping a steady stream of disposable adult labour through temporary immigration (the 'nanny chain') and outsourcing (the 'global assembly line'). This labour is cheap because it is embedded in a family economy where women, the elderly, and particularly children provide a host of services for free that families in the north must buy in the open market. Notwithstanding family wages, social security, childcare, free education and medical care, pensions and housing schemes, children are much less available for this kind of services in the north and tend therefore to become an increasingly prohibitive 'luxury'.

As an example that this is not the case in the south, one could take the discussion on 'informal safety nets' that followed on the critiques of IMF-led structural adjustment programmes. World Bank economists successfully claimed that where patterns of subsistence farming, mutual help in child care, the exchange of services and loans among neighbours and kin and so on are at work, in short where the family economy is still well in place as is widely the case in the global south, governments need not only provide additional welfare services but can safely 'do more with less' (Haddad, 1996; Ruel, 1999). What the economists failed to highlight was that children, who form anything between 50 and 70% of the population in the global south, play a crucial role in all these activities. In the same vein, the ILO brushes away:

> ... activities such as helping their parents care for the home and family, assisting in a family business or earning pocket money outside of school hours and during school holidays. (ILO, 2002, p. 15)

as not being 'child labour' but beneficial child work. Similarly, the World Bank believes that:

> Not all child labour is harmful. Many working children are within a stable and nurturing environment with their parents or under the protection of a guardian and can benefit in terms of socialisation and from informal education and training ... (Fallon, 1998, p. 5)

The problem then is not with the mundane, necessary work for life production, but with its preservation. It is in this key that one should read the policies put in place to deal with global child labour: as ways of naturalizing children's lifeworlds as sites where the extremely low-cost labour on which the new international political economy is premised would be unproblematically reproduced.

## Concluding Remarks

In the new world order that emerged after the fall of the Berlin Wall in 1989, the south was not merely allotted the role of provider of low-cost labour but, more importantly, of life producer. The centrality of life production lays bare how the family economy not only feeds market demands for people but also preserves and reproduces life itself over longer periods of time. What unfolded in the 1990s was a dramatically different representation of child labour than the historical one: from an internal social problem of nations states, it became an international relations issue dividing the world between the highly developed north and the low-income, low-labour standards but above all life producing south. The period of economic reform and market liberalisation of the 1990s brought in its wake a marked weakening of the autonomy of newly independent nations and growing northern intervention in their economic, political and social affairs. Images of suffering southern children multiplied, feeding a steady current of texts on the need to rescue and rehabilitate children apparently abandoned both by their own parents and their governments. Global child labour was part of this wider northern call to intervene in the south, but addressed particular constituencies and had specific repercussions on the division of labour between life producing and life consuming economic spheres.

For the vast majority of children in the south, global child labour reproduces the old colonial representation of the colonized as an endless source of labour power produced for free in the mythical moral economy of peasant families. But it also transforms this representation insofar that it disarticulates childhood from earlier territorial projects. The *limen* (Turner, 1970) or borderline between the global womb and global childhood is gradually shifting to new landscapes of symbolic power most clearly observable in the multiplication of mediatized child saving rituals. Of course globalisation also undermines this new world order, making it increasingly necessary for young men and women to seek supplementary sources of wealth through migration. But children are much less part of this movement and have remained overwhelmingly in the rural countryside or in the urban shantytowns. Demonstrating their ability to effectively deter autonomous children from leaving the global womb in order to invent childhood practices of their own, NGOs play a crucial role in peddling global childhood representations between reluctant communities and local elites, ritually staking out the mythical social order that embed the lifeworlds of today's children.

## Note

1. From the moment that children's protection is left to market forces, combating crime takes precedence above fulfilling even the most pressing social needs. It is yet too soon however to say if ILO Convention 182, as some critics have suggested, is in fact a legalisation of child labour, but what is clear is that failure to introduce core labour standards in the WTO means that violations of ILO Convention 182 cannot be the object of sanctions on equal footing with violations of WTO rules.

## References

Balagopalan, S. (2002) Constructing indigenous childhoods: colonialism, vocational training and the working child, *Childhood*, 9(1), 19–34.

Bey, M. (2003) The Mexican child, from work in the family to paid employment, *Childhood*, 10(3), 287–99.

Bradshaw, Y.W. (1993) New directions in international development: a focus on children, *Childhood*, 1, 134–42.

Bhukuth, Augendra (2005) Child labour and debt bondage: a case study of Brick Kiln workers in southeast India, *Journal of Asian and African Studies*, 40(4), 287–302.

Burgoon, B. and Wade, J. (2004) Patchwork solidarity: describing and explaining US and European labour internationalism, *Review of International Political Economy*, 11(5), 849–79.

Chirwa, Wiseman Chijere (1993) Child and youth labour on the Nyasaland plantations, 1890–1953, *Journal of Southern African Studies*, 19, 662–80.

Cornia, G., Jolly, R. and Stewart, F. (eds) (1987) *Adjustment with a Human Face, Vol. 1: Protecting the Vulnerable while Promoting Growth*, Oxford: Clarendon.

Couldry, N. (2003) *Media Rituals, A Critical Approach*, London: Routledge.

Cunningham, H. (1991) *The Children of the Poor: Representations of Childhood since the Seventeenth Century*, Cambridge, MA: Blackwell.

De Suremain, C.-E. (2000) Coffee beans and the seeds of labour: Child labour in Guatemalan Plantations, in Schlemmer (ed.) *The Exploited Child*, London and New York: ZED, 231–38.

Escobar, A. (1995) *Encountering Development, The Making and Unmaking of the Third World*, Princeton, NJ: Princeton University Press.

Fallon, P. and Tzannatos, Z. (1998) *Child Labour, Issues and Directions for the World Bank*, Washington: The World Bank, Social Protection, Human Development Network.

Fyfe, A. (2001) Child labour and education, Revisiting the policy debate, in K. Lieten and B. White (eds) *Child Labour, Policy Options*, Amsterdam: Aksant, 67–84.

Gellner E. (1983) *Nations and Nationalism*, Ithaca, NY: Cornell UP.

Greer, B. (1994) Invisible hands: the political economy of child labour in colonial Zimbabwe, 1890–1930, *Journal of Southern African Studies*, 20(1), 27–52.

Haddad, L. and Zeller, M. (1996) How can safety nets do more with less? General Issues with Some evidence from Southern Africa, Washington: International Food Policy Institute, FCND Discussion paper No 16.

Hearn, Julie (2002) The 'invisible' NGO: US evangelical missions in Kenya, *Journal of Religion in Africa*, 32(1), 32–60.

Hecht, T. (1998) *At Home in the Street, Street Children of Northeast Brazil*, Cambridge: Cambridge University Press.

ICFTU, Campaigning for an end to child labour, 17th World Congress of the ICFTU, Durban, 3–7 April 2000.

ILO (2002) A Future without child labour, Global report under the follow-up to the ILO Declaration on Fundamental Principles and Rights at Work, Geneva: ILO.

Invernizzi, A. (2001) *La vie quotidienne des enfants travailleurs, Stratégies de survie et socialisation dans les rues de Lima*, Paris: L'Harmattan.

Kenny, Mary Lorena (1999) No visible means of support: child labor in urban northeast Brazil, *Human Organization*, 58(4), 375–86.

Koshy, S. (1999) From Cold War to Trade War: Neocolonialism and Human Rights, *Social Text*, 58, 1–32.

Lange, M.F. (2000) The demand for labour within the household: Child labour in Togo, in B. Schlemmer (ed.) *The Exploited Child*, London: ZED.

Lavallette, M. (1994). *Child Employment in the Capitalist Labour Market*, Aldershot: Avebury.

Levine, S. (1999) Bittersweet harvest, children, work and the global march against child labour in the post-apartheid state, *Critique of Anthropology*, 19(2), 139–55.

Libal, K. (2002) Realizing modernity through the robust Turkish child, 1923–1938, in D.T. Cook (ed.) *Symbolic Childhood*, New York: Peter Lang, 89–108.

Mamdani, M. (1996) *Citizen and Subject, Contemporary Africa and the Legacy of Late Colonialism*, Princeton, NJ, Princeton University Press.

Michener, V.J. (1998) The participatory approach: Contradiction and Co-option in Burkina Faso, *World Development*, 26(12), 2105–18.

Montgomery, H. (2002) *Modern Babylon? Prostituting Children in Thailand*, New York: Berghahn Books.

Myrstad, G. (1999) What can trade unions do to combat child labour? *Childhood*, 6(1), 75–88.

Nieuwenhuys, O. (1994) *Children's Lifeworlds, Gender, Welfare and Labour in the Developing World*, London: Routlegde.

Nieuwenhuys O. (2001) By the sweat of their brow? Street children, NGOs and children's rights in Addis Ababa, *Africa*, 71(4) 539–57.

Polanyi, K. (2001 [1941]) *The Great Transformation, The Political and Economic Origins of our Time*, Boston, MA: Beacon Press.

Pupavac, V. (2001) Mysanthropy without borders: The international children's rights regime, *Disasters*, 25(2), 95–112.

Rahikainen, M. (2001) Historical and present-day child labour: is there a gap or a bridge between them? *Continuity and Change*, 16, 137–56.

Rahikanen, M. (2004) *Centuries of Child Labour, European Experiences from the Seventeenth to the Twentieth Century*, Aldershot: Ashgate Studies in Labour History.

Reynolds, P. (1991) *Dance Civet Cat. Child Labour in the Zambezi Valley*, London: ZED.

Ramanathan, U. (2000) The public policy problem: labour and the law in India, in B. Schlemmer (ed.) *The Exploited Child*, London: ZED, 146–59.

Rist, G. (2001) *The History of Development, From Western belief to Global Faith*, London: ZED.

Robson, E. (2004) Children at work in rural northern Nigeria: patterns of age, space and gender, *Journal of Rural Studies*, 20, 193–210.

Ruel, M.T., Haddad, L. and Garrett, J.L. (1999) Some urban facts of life: implications for research and policy, *World Development*, 27(11), 1917–38.

Scheuerman, W. (2001) False humanitarianism? US advocacy of transnational labour protection, *Review of International Political Economy*, 8(3), 359–88.

Schildkrout, E. (2002) Socio-economic roles of children in urban Kano, *Childhood*, 9(3), 344–68.

Simelane, Hamilton Sipho (1998) Landlords, the State and child labor in colonial Swaziland, 1914–1947, *The International Journal of African Historical Studies*, 31(3), 571–93.

Swaminathan, M. (1998) Economic growth and the persistence of child labour: evidence from an Indian city, *World Development*, 26(8), 1513–28.

Taracena, E. (2003) A schooling model for working children in Mexico: The case of children of Indian origin working as agricultural workers during the harvest, *Childhood*, 10(3), 301–18.

Turner, V. (1970) *The Forest of Symbols, Aspects of Ndembu Ritual*, Ithaca, NY: Cornell UP.

Tvedt, T. (2002) Development NGOs: actors in a global civil society or in a new international social system?, *Voluntas: International Journal of Voluntary and Nonprofit Organizations*, 13(4), 363–75.

Ulvila, M. and Hossain, F. (2002) Development NGOs and political participation of the poor in Bangladesh and Nepal, *Voluntas: International Journal of Voluntary and Nonprofit Organisations*, 13(2), 149–63.

Van Haer, N. (1982) Child labour and the development of capitalist agriculture in Ghana, *Development and Change*, 13, 499–514.

Weiner, M. (1991) *The Child and the State in India*, Princeton, NJ: Princeton University Press.

White S.C. (1999) Ngos, civil society and the state in Bangladesh: the politics of representing the poor, *Development and Change*, 30, 307–26.

White, B. (2001) Childhood, work and education, 1900–2000: The Netherlands and Netherlands Indies/ Indonesia Compared, *Brood en Rozen, Tijdschrift voor de Geschiedenis van Sociale Bewegingen*, 6(4), 105–19.

Wolf, E.R. (1982) *Europe and the People without History*, Berkeley, CA: University of California Press.

World Bank (2001) *World Development Report*, Washington: The World Bank.

# Children, Young People, UNICEF and Participation

TRACEY SKELTON

*Department of Geography, Loughborough University, Ashby Road, Loughborough LE11 3TU, UK*

## Introduction

'Participation' appears to be *the* word, concept and discourse to engage with when doing research or working with children and young people in the context of development. It is almost held up as the panacea for all the problems young people and children face in the South (and it is gaining precedence in the North too, see Mayo, 2001). However, as with all prevailing and encompassing discourses it is important that it is closely interrogated, unpacked and critiqued. This article contributes to the growing literature which engages with the highly complex and dynamic discourse and practice of children's and young people's participation.

This paper will initially examine the UN Convention on the Rights of the Child (CRC) which has been identified as the trigger to a revolutionary step in the recognition of children as 'human beings' rather than has 'human becomings' (Kjørholt, 2001, 2004). It will then provide a critical commentary on selected aspects of wider discussions which have contributed to a reconstitution of the concept of childhood, children and social participation. The next part of the article will take one specific published case study through which to explore a particular interpretation of children's and young people's 'participation'; UNICEF's State of the World's Children 2003 (hereafter SOWC) report published in 2002. This allows us to examine the ways in which one of the key initiators and sponsors of the CRC, UNICEF, constructs the concepts of childhood and participation in relation to social development. Hence it is a way to see a specific part of the UN at work in the way it practices its own doctrine of participation. Throughout the discussion of the SOWC this paper will provide a critical discourse analysis of 'participation' and the construction of 'childhood'

At this stage I wish to clarify this paper's definition of 'children', 'young people' and 'childhood'. It can be quite hard to define children and young people (Skelton, 2000; Weller, 2006). Definitions of who is a child and who might be called a young person often vary over time, space and also within a particular context whether the focus is social, economic, political, legal or cultural. Hence, within the UK children/young people are legally bound to remain in school until they are 16, can work part-time for a number of hours per week from the age of 14, can learn to drive when they are 17 and are entitled to vote when they are 18. Hence different rights are dependent on different ages. There is often far too much emphasis placed on age and yet in most cases this is expedient, even if problematic; age categorisations tend to be used despite their problems. Nevertheless, this can be culturally problematic because different societies have different definitions but these seem to be being eroded by a global definition of 'the child'.[1]

UNESCO (2002) talks of young people being aged 15–24. The UNCRC defines children as 'every human being below the age of eighteen years unless under the law applicable to the child, majority is attained earlier' (UN CRC, 1989, Part 1, Article 1). Hence, a particular country can change the definition of child by changing the age of majority—but at this stage children move into the realms of adult rights which are often very different to those afforded children. This definition and the 'magical age of 18' is used to define children and appears, by default to, include young people; certainly within UNICEF and UNESCO definitions. Young people are not very often mentioned *per se* in documentation; indeed, throughout the UNICEF document examined here, the terms 'child' and 'children' are prevalent, 'young people', 'adolescent' are less so. In contrast, in a UNESCO (2002) document the term 'child' is not used even though those under 18 are recognised within the document and according to the CRC are still classified as children. However, if we consider these two international organisations (UNICEF and UNESCO) and the definitions embedded within their documents it can be interpreted from the discourse that 'child' could be a term inclusive of young people. This point is an important one to raise because it illustrates one of the problems of the predominance of discourses on children which do not even mention young people. Young people who do not consider themselves to be children anymore, and are in many ways not perceived as such by wider society, may feel that the rights defined for children do not apply to them. They can be in a kind of limbo between childhood and adulthood (Sibley, 1995). They are certainly rarely mentioned in global and national discourses of 'the child'.[2] Where this article uses the terms 'child' and 'childhood' (because this is the predominant terminology within the UNICEF material under focus) it also includes, and sometimes explicitly names, young people. This is because the author feels it is important that readers are reminded that young people, although in a sort of liminal, inbetween space, play an important role in society and it is important that their presence is included. The concept of 'childhood' is much harder to define, and consequently in the subsequent section of this paper there is a brief discussion, well rehearsed in wider 'childhood' studies, of what are the current perceptions of 'childhood' and how these can be explicitly related to discussions about participation.

## Children, Childhood and Participation

The notion of children and young people being encouraged and facilitated to participate in a range of social and cultural contexts and institutions, particularly where these have a direct impact on their lives, has grown in significance throughout the 1990s (Hart, 1997; Johnson *et al.*, 1998; Matthews *et al.*, 1999; Kjørholt, 2001, 2002, 2004; Auriat *et al.*, 2001; Barrow, 2002; Chawla, 2002b; UNESCO, 2002; Freeman *et al.*, 2003; Kelley, 2006; McNeish and Gill, 2006). Many acknowledge that this focus on participation is strongly connected with the introduction of the Convention of the Rights of the Child (Matthews *et al.*, 1999; Barrow, 2001; Chawla, 2002b; Freeman *et al.*, 2003) The CRC was adopted and opened for signature, ratification and accession by General Assembly resolution 44/25 on November 20th, 1989. It came into force on September 2nd in 1990 and has been ratified by all but two countries of the 193 UN member states (those two being Somalia and the USA) throughout the early 1990s. The UK Government signed it in 1991 (Matthews, 2005).

The CRC of 1989 marked a change in approach to the world's children. In 1959, through UNICEF, the UN General Assembly adopted the Declaration of the Rights of the Child which defined children's rights to protection, education, health care, shelter and good nutrition. Hence, in the early stages, goals tended to be about the practical support and care of children, especially at times of crisis. The emphasis at this time was on provision and protection. The CRC restated the first two 'Ps'—*provision* and *protection* but it also introduced a third 'P'—*participation*. A range of commentators have argued that this has been its most radical contribution and has been part of a paradigmatic shift in thinking about children in global and national contexts (Hammarberg, 1990; Freeman, 1992; Cantwell, 1993; Kjørholt, 2001; Mayo, 2001; Chawla, 2002b). Focusing on children's rights in the Caribbean, Christine Barrow (2002, p. xv) argues:

> The distinguishing feature of the CRC lies in its emphasis on the *participation* of children in decisions affecting their own lives. Previous declarations had adopted a restricted and paternalistic view of children ... The CRC, in contrast, acknowledges the autonomy of the child and the accompanying principles of social inclusion, self-determination and empowerment.[3]

It is important, in order to demonstrate the significant emphasis placed on participation, that we take a close look at the CRC (and other Agendas and documents) in order to identify the ways in which participation is woven throughout the convention. 'Participation' is not a stand-alone article but is *embedded* within the CRC and, importantly, it is also evident in other international agreements where children *per se* were not necessarily the central focus. Hence the discourse of 'participation' spreads further than the CRC but it is the CRC that clearly defines the importance of children's participation.

The significant articles of the CRC are those from 12.1 to 15.1 (Auriat *et al.*, 2001; Chawla, 2002b). They state that children who are capable of forming their own views should have the right to express them freely in all matters which affect them; that children should have the right to freedom of expression and access to information; that children should have rights to freedom of thought, conscience and religion; that children should have the right to freedom of association and peaceful assembly (UNICEF, 1990). Louise Chawla (2002b, p. 26) argues that there are other articles are relevant to the principle of participation, specifically in relation to the quality of children's environments, namely articles 27, 29, 30 and 31.

In the document '*UNESCO-mainstreaming the needs of youth*' (2002) it is argued that 'the most ardent wish of young people is to *participate*, as full and equal citizens, in today's world' (pp. 2, emphasis in the original). On page 4 the document states: 'Today, the Organization [UNESCO] is mobilizing support from Member States and other partners (non-Governmental organizations (NGOs), civil society organizations and associations) and is moving even further to integrate youth in terms of participation, partnership and empowerment'. Hence there is an explicit focus on the agents (other than young people themselves) who need to work towards the full participation of young people.

The United Nations Conference on Environment and Development in 1992 produced Agenda 21 and a Programme of Action. There was a chapter on children and young people stating that they are a major group who should be part of participatory processes for sustainable development and environmental improvement. Similarly the Habitat Agenda from the United Nations Conference on Human Settlements in 1996 states that governments must utilise participatory approaches which include intergenerational interests relating to sustainable human settlements demanding special attention to children (Chawla, 2002b, p. 27).

Mayo (2001) states that children's participation has entered the mainstream vocabulary of development but that the practice lags behind the rhetoric. She argues that when there is children's participation it is often more about 'tokenism' or 'decoration' (Hart, 1997) and that children's voices are still not being listened to. However, there are clear examples of very effective participation on the part of children evident in the eight country programme *Growing up in Cities* (Chawla, 2002a). Nevertheless, despite the programme's success in children's participation the study 'found a large gap between the rhetoric of international agreements and the reality of authorities' provisions for children' (Chawla, 2002b, p. 31).

Of course the CRC, and its introduction of the concept of participation for children, didn't appear out of a vacuum. There was a range of groups (NGOs, campaigning groups focusing on education, child labour, abuse issues, etc.) who were pushing for participation as a feature of children's rights. In particular countries individual changes to rights legislation for children were being introduced in the late 1980s and throughout the early 1990s. In some cases this was as a response to specific events or practices in which the lack of acceptance of children as being protagonists or as having the right to be listened to had had dire consequences for children. In the UK the 1989 Children's Act of England and Wales was in part a response to a large number of cases of systematic child abuse that had not been properly investigated or dealt with effectively.

Alongside changes in legal and political debates relating to children, and the growing development discourse around participation, there was also a conceptual shift taking place within social studies of childhood and children (James and Prout, 1990; Qvortrup, 1994; James *et al.*, 1998). Over the past 15–20 years there has been a paradigmatic shift in the social studies of childhood which has critically engaged with ideas about what constitutes 'childhood' and which also uses new methodological approaches such as discourse analysis (Kjørholt, 2004). In some ways the new conceptualisations of children and childhood has enabled the more 'practice-based' discussions about participation to emerge.

The 'new' social studies of childhood have been discussed elsewhere within geography (cf. Holloway and Valentine, 2000, for an excellent summary[4]). All I wish to do here is pull out *some* of the key features of this 'new' approach that frames the critical discourse analysis of this particular paper and examine what roles these play in the conceptualisation of participation.

• Childhood is understood to be a social rather than a natural phenomenon. As a social construct it is part of particular social, historical and cultural contexts. Alongside the conceptualisation of gender, childhood is now recognised not to be biologically determined nor a constant. If childhood is a social construction then its meaning and practice can change over time. What childhood has meant to the United Nations has changed over time from the 1959 Declaration through to the CRC of 1989 (as discussed above). The possibility of social change in the conceptualisation of children and young people is one facet that has allowed for the recognition of their participatory potential.

- Children are seen as social actors and agents. This transforms previous conceptualisations which saw children as passive beneficiaries, silent objects of concern who are dependent on adult control and care (Barrow, 2002). It is important to stress here that not all conceptualisations of children have gone through such a transformation. There is still a pervasive perception of children as dependent, in need of protection and unable to adequately articulate their own needs.[5] Children are understood to be competent and so entitled to have the right to participate in society and have a say in issues which affect their lives. They are constructed as 'being' instead of 'becoming' (Qvortrup, 1994, this is a conceptual point that I return to later in the paper).
- There are multiple childhoods which are constructed, in part by children themselves, which are dynamic across time and space. There are therefore different cultural interpretations of childhood but these are currently locked in engagements (even battles) with universalised global discourses on childhood which hold certain principles about children's rights as central. As Jo Boyden points out 'highly selective, stereotyped perceptions of childhood . . . have been exported from the industrial world to the South. They have provided a focal point for the development of both human rights legislation at the international level and social policy at the national level in a wide range of countries' (1990, p. 191). I return to the problems of this 'northern-centrism' in global definitions of childhood later in the paper. What Boyden's work shows very clearly is that while the new social studies of childhood recognise the cultural and spatial differentiations for the meaning of childhood, international agreements appear to be not so well informed.
- If children are seen as competent social actors then they have the right and the ability to act as participants. This is a participation which goes beyond the fact that 'life itself presupposes participation' (Kjørholt, 2004, p. 3). However, in any context, how, where and when children should participate is a cultural construction. Their participation thence constructs a particular social practice which in turn can establish the scale and form of children's participation.

Consequently the recent perspectives on children's participation have emerged from a range of fields: global discourses and conventions; individual countries and state legislative practices; development discourses and actors; academic paradigm shifts; and children and young people themselves.

There are though extensive problematics related to children's participation, some of which this paper illustrates. There are anxieties about whether children's participation can be genuine and effective or rather simply symbolic, a form of window dressing which acts as tokenism or decoration (Hart, 1997). There are serious concerns about what children and young people themselves feel about their participation and whether they are being listened to effectively (Stephens, 1995; Mayo, 2001; Kjørholt, 2002; Percy-Smith, 2002). As Nelson and Wright have argued (1995, p. 2) participation is a 'warmly persuasive word' that 'seems never to be used unfavourably and is never given any positive opposing or distinguishing term'. Much of the discussion and practice of participation is based upon universalised and normative assumptions. There is a universal view of children and childhood (which many argue is a based upon a Northern discourse, Boyden, 1990) and also a presumption of the value of such participation. In view of the predominance of the notion of children's participation, and the fact that it has become a highly persuasive global discourse, it is essential that there is a continual critical interrogation of what it means and whether its practice is beneficial or detrimental for the children involved and children more generally. This article is part of that critical view.

**UNICEF**

The UN created the United Nations Children's Fund (UNICEF) in December 1946. It was specifically set up to help children who were facing famine and disease in Europe after the Second World War. UNICEF then led the way in the establishment of the 1959 UN Declaration of the Rights of the Child which established the centrality of provision and protection for children. More recently it has been a key protagonist for the writing and signing of the Convention on the Rights of the Child (CRC) in 1989 and in continuing to ensure the CRC is put into practice. It is a central player in the UN and ensures that children and young people are part of the agenda. An example of this is its instigation of the United Nations' Special Session on Children in May 2002 (New York).

UNICEF is a major player in development which specifically relates to children, families and education (Ansell, 2005). It interacts with, and has to try and counter, some of the most damaging processes connected with globalisation (see Katz, 2004, for an excellent analysis of the role global processes play in children's lives). In addition, it is dedicated to the healthy and sustainable reproduction of future generations. As a global player it structures dominant discourses of the 'global child' and constructs children, young people and childhood in particular ways. Finally, UNICEF, as part of the UN, is committed to the principle of effective democracy and 'authentic participation'.[6] It supports a variety of schemes within nations (as well as regions) that focus on children's participation in democratic processes and consequently contributes to the construction of national childhoods. It can also encourage agenda setting within regions and states (see below in relation to the Organisation for Security and Co-operation in Europe).

The CRC has challenged and developed the ways in which development agencies, national governments and NGOs work and 'do' development *to* and/or *with* children.[7] Indeed, UNICEF has a sub-title evident on most of its web pages and reports which reflects its work and goals:

> *For every child*
> *health, education, equality, protection*
> *ADVANCE HUMANITY*

The element of progress and a focus on the future is captured in the word 'advance'. We might critique this as a link with 'modernisation' development theory and the expectation of constant improvement. We could also applaud the sentiment which appears inclusive of everyone and states some of the elements that are essential for child welfare.

Early in the twenty-first century, UNICEF has organised several major global events to raise the profile of children and young people. In 2001 it organised the worldwide campaign '*Say 'Yes' For Children*' beginning in March. Adults and children were asked to make a pledge by saying 'yes' to the following statement: 'I believe that all children should be free to grow in health, peace and dignity'. Ninety-five million pledges (including 20 million from China and 16 million from Turkey) were presented at the Children's Forum in New York in 2002. As part of this campaign children were asked to list three priorities. The top three issues were, education, discrimination and poverty. The Forum then reported to the Special Session (May 2002, see below).

In the 12 months prior to the Special Session for Children of the UN General Assembly (May 2002) 40,000 children aged 9–18 in 72 countries answered a poll about their everyday realities and their hopes for the future. The central thematic was '*a world fit for children*'. These survey data have informed a range of subsequent initiatives, including one by the Organisation for Security and Co-operation in Europe. The OSCE is the largest security organisation in the world and it has asked its field missions to use the survey results 'to inform and guide their programmes aimed at strengthening democratic citizenship, civic education, conflict prevention and security' (SOWC, 2003, p. 51).

In May 2002 the General Assembly of the UN held a special session on children to exclusively discuss children's issues, for the first time. Also, for the first time, children were included as official delegates representing governments and NGOs. All of these events have a played a role in the UNICEF goal of children's participation which is then summarised in The State of the World's Children 2003 report. The year 2002 was therefore a key time in the history of UNICEF and the UN in their approach towards children and young people. In particular, 2002 can provide an insight into the ways in which UNICEF (and the UN) is putting the CRC understanding of children's participation into practice.

It is clear, therefore, that UNICEF has broadened its scope substantially while keeping the same early goals as part of its *raison d'être*. The centrality of education and nutrition are evident in the list below of the key foci of past 'State of the World's Children' reports. In addition, there are other themes which occupy the time and energy of this vast global organisation which funds, supports and endorses an enormous number of development initiatives both at times of crisis and throughout the more mundane everydayness of life for children and young people.

---

**The State of the World's Children: themes over time**

- Children in War 1996
- Child Labour 1997
- Nutrition 1998
- Education 1999
- Generic call to re-affirm promises made to children at the 1990 World Summit for Children 2000
- Early Childhood 2001
- Leadership 2002
- Child Participation 2003
- Girls' Education and Development 2004
- Childhood Under Threat 2005

---

There are, therefore, specific themes which are deemed as central to UNICEF's role in improving the lives of children, young people and families and these establish a global discourse around childhood. Within individual countries there are deep impacts. The SOWC reports both set an agenda and report. Hence in any particular country specific priorities in relation to children and young people can be determined by the focus of UNICEF as it can have a direct influence through its many field officers and its funding of research and particular projects. This article will now explore one particular moment and product of UNICEF's global discourse and critically engage with UNICEF's own report dedicated to *participation*.

## The State of the World's Children 2003 Report: Child Participation

This report (available in PDF format on the UNICEF web site) is broken into nine chapters, interspersed with eight panels. The chapters tackle specific questions and areas relating to participation but in a somewhat generic way. The panels illustrate (in some cases literally with photographs and drawings by children) more specific examples of children's and young people's participation. The report is written in a highly accessible way, not least to make sure older children can engage with it. There are three maps that visually represent the results of the special poll conducted in 2001 in 72 countries (see above). Pages 81–119 provide nine tables which present the economic and social statistics on the countries and territories of the world which have a particular reference to children's well being. This article engages with selected elements within the narrative sections of the report, particularly in the chapters.

First, it is important to say something about the specific selection of this one report and the methodological approach used. The 2003 report is the one UNICEF report dedicated to the principle of participation. It is a report on various events that took place under the auspices of the UN in 2001 2002 and in which children's and young people's participation was an integrated part and which have been identified above. At the end of the General Assembly there was a pledge made to build 'a world fit for children' and world leaders stated a commitment to change the world *with* children's participation. Hence the 2003 report is the public statement by the UN of its own engagement with the practice of participation. In chapter 2 it asked the question—*Why participation, why now?* (the 'answers' to this question are discussed below) The fact that this particular report explicitly raises this question means that it is an important document which forms part of the global discourse about children's and young people's participation and hence is central to a critical review of such discourses.

Above I have stated that I used a critical discourse analysis of this particular publication. Discourse analysis aims to explore the outcomes of discourse in terms of actions, perceptions, or attitudes; to identify the regulatory frameworks within which groups of statements are produced, circulated and communicated; and to uncover the support mechanisms that uphold certain structures and rules over statements about people, events, places as unchallengeable, 'normal' or 'common-sense' (paraphrased from Waitt, 2005, pp. 164–5). Hence in this discourse analysis the aim is to examine the actions, perceptions and attitudes as they are presented by the UN about its own practices in relation to children's participation. By examining a 'global document' with a high level of circulation it is possible to say something about the regulatory framework within the report is both produced and the framework it functions to establish. There is also the opportunity to explore the ways in which the UN establishes the principle of certain types of children's participation as 'normal' and as 'common-sense'. Critically examining the discourse of the 2003 SOWC report means that it is possible to offer a critical commentary on the UN's practice in the field of participation and cast a vigilant gaze over a product that forms part of an 'accepted' global message.

Methodologically I drew upon Rose's (2001) and Waitt's (2005) strategies of how to approach discourse analysis. When I began working on this article it was the first time I had read a SOWC report hence it was easy for me to approach the text without any pre-existing categories and with an open mind. I read the text repeatedly—twice without making any notes or comments so that I could have a grasp of the text as a whole and understanding something about the different features of this 124-page text (as outlined above). On the third reading coding was carried out to identify any repeated or dominant themes; these were highlighted on the text and noted extensively and separately. The ways in which certain themes, language styles and images were used to persuade and convey a particular commentary were recorded as were any inconsistencies and contradictions within the report. Throughout the discourse analysis questions were asked about what was not present in the text and details which echoed or contradicted debates about childhood and participation were recorded.

The report is extremely rich and could not possible be reported in all the detail that was analysed. What I present here are *selected* thematics/concepts combined with a critical commentary upon, and questioning of, them. There is a great deal in this report that we can praise. I have been impressed with the ways in which UNICEF has taken specific action to ensure children's and young people's participation through its own projects and structures. Much of what is argued for here in the report I politically and intellectually would endorse and support. There are many echoes in this report of the 'new' conceptualisations of children discussed above as: social agents; deserving of rights; being competent at responsibilities; and important players in contemporary society. However, in view of the wide-reaching impact UNICEF and its discourse has in a global context, it is important to maintain a critical eye on the practice and engage with the possible hidden meanings, the 'regimes of truth', behind the rhetoric (Foucault, 1980; Skelton, 1999; Waitt, 2005).

## *Children as Hope/Children as Innocence: Problematic Representations?*

The first page of the document is a foreword statement from Kofi Annan, secretary-general of the UN. He argues that the report is true to the spirit of the General Assembly Special Session on Children (2002). He states that the presence of children 'transformed the atmosphere of the United Nations' because the children 'introduced their passions, questions, fears, challenges, enthusiasm and optimism. They brought us their ideas, hopes and dreams'.

This notion of children as the holders of dreams and hopes is a recurrent trope through the report. While it is a positive representation of children there is also a possible sub-text which renders children as innocent, unsullied and unspoiled by the world at large (a reflection of previous discourses of children which have linked them to nature and of childhood as a state of innocence.) As Boyden (1990) notes this is very much a Northern/industrial nations' representation of childhood. It can smack of the preternatural child which wanders/wonders wide-eyed into the world with a fresh, but possibly, naive perspective. Such a discourse is made visible through the use of images. Fresh-faced, smiling, laughing, curious children (often photographed by other children, but selected for the report by adults) gaze out of the pages. The frank, honest and straightforward statements of children made at the special session for children or at a youth parliament meeting are writ large to demonstrate the directness of children's and young people's demands. There is a sense of children 'saying it as it is'. However, such statements can be defined as simplistic and devoid of the complexities of 'real life' which adults presume they understand so much better. The statements and the images are inspirational and moving—but is this a form of manipulation? Bombarded with images of children to invoke sympathy, concern, even pity, can we detach ourselves from the similar images used in reports about how young people should be participants in their own futures? The representation of children is also a highly problematic issue. Images of children can be read in so many different ways and they are a pervasive iconography of the global world. Is it possible that images of children might be used as a substitute for the presence, in an active, everyday way, of the children themselves?

## *A Concept of Participation: 'Authentic' Participation*

On page 3 the report argues that 'children have always participated in life', and many have made significant differences in their life-worlds. Childhood as a social construct has been transformed and consequently children as a group are recognising that they have, and are being recognised as having, rights and social agency. The definition for participation is taken from Roger Hart's essay *Children's Participation: from tokenism to citizenship* (1992, p. 5):

> The process of sharing decisions which affect one's life and the life of the community in which one lives. It is the means by which democracy is built and it is a standard against which democracies should be measured.

The emphasis is placed upon meaningful, genuine, authentic participation. Without caution children's participation can amount to manipulation, decoration or tokenism (SOWC, 2003, p. 5). Children's participation can often become more about the adults than about the young people. It can even be repressive, exploitative or abusive. UNICEF contrasts this with 'authentic participation' which

> must start from children and young people themselves, on their own terms, within their own realities and in pursuit of their own visions, dreams, hopes and concerns.[8] Children need information, support and favourable conditions in order to participate appropriately and in a way that enhances their dignity and self-esteem. (SOWC, 2003, p. 5).

The report later acknowledges that to bring about authentic participation for children demands a radical shift in adult thinking and behaviour. Adults have to recognise that they can no longer define the world on their terms only.

The problem with such notions of participation is that it is decontextualised. Apart from one sentence 'it also depends on the given sociocultural, economic and political context' (SOWC, 2003, p. 5) there is no other recognition that children have very different means through which they can (or cannot) express their hopes and dreams. It might be the case that that poorer children lack the social capital and education required to be able to even begin to articulate their wishes to themselves or their peers let alone to begin to participate alongside adults. As the UNESCO Growing up in Cities project has shown children from poor backgrounds do have a great deal to say that proves extremely useful *if* adults are willing to *ask* the questions and *listen* to the answers in ways that facilitate and encourage their participation (Chawla, 2002a). Hence there have to be appropriate settings, practices and approaches to ensure that all types of children can have their say. While UNICEF clearly does a great deal of work to try and rectify inequalities between children the deep impact poverty may have on children's (and adults') abilities to participate 'authentically' is not made explicit in the document. Chapter 5, *The Sharpest Edge*, does mention the ways in which adolescents and children face exploitation and abuse by adults. It is just three and a half pages long and the role of poverty is not made explicit.[9]

Hence, although there is a hegemonic discourse around children as social participants they are still found in marginal positions (Boyden and Holden, 1991; Dodman, 2004; Kjørholt, 2004). The focus on children as individuals is potentially important as part of debates about their rights but in poorer communities separating children from inter-generational networks in their communities can have disastrous consequences (Boyden, 1990). Such networks are often part of complex reciprocal relationships that are invaluable at times of crisis or insecurity, especially for individual families. Poverty impacts on many childhoods and is part of their lived context of the type and extent of their participation. Indeed in poor communities children may be vibrant participants whether through their involvement in paid work, labour within their family or even as a source of amusement and laughter for weary adults. Is this any the less 'authentic' because it does not take place in a public arena and is not about children identifying their own needs as somehow individualised and separate? The context of children's participation is often a missing part of the discussion and consequently it is such contexts which require further research in order to identify the means to effective and meaningful participation.

### The Theme of Democracy: Why Children 'Must' Participate

Chapter two of the report raises the question of why participation is important and why this should be now. It recognises the criticism which may be levelled against the call which is that many segments of the world's population are denied participation so is not including children just going too far? Children are dying in their thousands everyday so why is listening to their voices so important? The chapter then offers a range of 'because' explanations.

A key 'because' relates to democracy and this is a theme which emerges throughout the rest of the report. It is directly linked with development as a process. It is argued that democracy is important in the process of gaining international peace and development. Democracy has certain core values that include respect for, and the dignity of, all people.[10] These values, and the right to participate in order to secure such values, are best learned in childhood. In this sense 'participation is a keystone for cohesive societies, which, in turn, are the keystone for peace in the world' (SOWC, 2003, p. 10).

Later a sub-heading reiterates the linkage; *'Democracy begins with children'*. As world leaders are faced with greater and greater threats from terrorism, extreme poverty and people who feel increasingly disenfranchised they are turning to processes of 'deepening democracy'; a democracy which is more inclusive and responsive. To this end Member States of the UN pledged in the Millennium Declaration that they would 'spare no effort to promote democracy and strengthen the rule of law, as well as respect for all internationally recognized human rights and fundamental freedoms, including the right to development' (SOWC, 2003, pp. 11–12). Additionally there was a promise to meet the Millennium Development Goals by 2015, and six of the eight are related to children. This links through to the Special Session on Children where governments committed to protect the rights of children and also safeguard their well being through other development processes.

The explicit link between democracy, children and development is then claimed (SOWC, 2003, p. 12) and, I would argue, established as a 'regime of truth'. If government, national agencies and other organisations do not vouchsafe children's rights and well being then the development goals will never be met. Poverty will remain and 'democracy will surely wither' (*ibid.*). This might be judged a tautology. If six of the eight Millennium goals relate to children, then, not to do what the goals list for children will automatically mean that they fail. However, the rhetoric serves its purpose. Children's rights, democracy and successful development have been firmly established as a dominant discourse.

The next subheading, *'The needs of democracy'*, then tells the story from a slightly different perspective. The disenchantment young people have towards democratic processes is identified as causing the greatest concern. However, rather than see this as a message of the problematic nature of contemporary democratic processes, UNICEF ploughs on with a commitment to democracy and a determination to get children involved. It would seem it is politic to listen to young people's voices only so far. When they are clearly critical of a particular practice/institution it is not about listening but more about getting them to participate in a system they have already expressed concern about. This is an inconsistency between the rhetoric and the practice of participation in relation to young people.

Having established that democracy needs young people to participate if it is to have a future, the final section of Chapter 2 is titled *'The hope for democracy'* (no doubt the term 'hope' is deliberate). The emphasis here is on the way in which democratic citizenship and understanding can be promoted. The best way, it is argued, is to begin such promotion in early childhood: 'the place to start to build democracy is with children—from what they learn in the process of their growth and development' (SOWC, 2003, p. 13).

This process of acquiring an understanding of 'democratic citizenship' is linked to participation. Children who are encouraged to participate in their families, schools, communities and societies are described as being more self-confident, more aware of what is happening around them and better behaved and respectful.

Hence a new generation of caring, responsible, aware and hard working citizens are to be developed. However, isn't this also about conformity, duty, and an acceptance of the status quo with perhaps some limited expectation of change, a replication of the values of adults? There is little room in this model of citizenship for creativity, disruption, rebellion or resistance.[11] Can young people and children genuinely gain from participating in a system, a process, a model for development, that previously has consistently marginalised them?

## *Citizens of the Future? Beings of the Now?*

The concept of democratic citizenship has been considered above. What I want to do in this final section of this paper is consider the ways in which the UNICEF report's representation of citizenship connects with the somewhat contradictory construction of children as 'becoming' or as 'being'.

In the first chapter *'Children must be heard'* there is a direct engagement with the anxiety adults express about children's participation undermining adult authority within the family and society (SOWC, 2003, p. 4). UNICEF counters that '[t]he social give and take of participation encourages children to assume increasing responsibilities as active, tolerant and democratic citizens *in formation*' (*ibid.:* my emphasis). This means that children are understood to be in the state of 'becoming'.[12] They have a future ahead of them and participation is a means to enable that future to be positive for them as they grow/develop into the 'right kind' of citizens.

The 'becoming' element of childhood is reiterated in two of the five 'because' reasons for children's participation in chapter 2 (SOWC, 2003, p. 9). '*Because* promoting meaningful and quality participation of children and adolescents is essential in ensuring their growth and development' [emphasis in original]. Another 'because' states that 'authentic and meaningful participation prepares children for their stake in the future'. In each case children are assumed to be in process, to be 'becoming' something else (something better, something more valuable?).[13]

Interestingly, within the same report are direct quotes from young people and children who were present and spoke at the major meetings organised and discussed above. These quotes by the children themselves firmly establish their sense of self as 'beings'. These are three particularly clear examples:

> 'In 1990 our countries signed the CRC but they have done next to nothing to realize it', said a 17-year-old delegate, his body shaking as he spoke, though out of sheer passion rather than nerves. 'We agree with your promises but *now* you have to show you mean it. I am talking from the heart—you must do the same'. (pp. 61, my emphasis)

> We the children are experts on *being* 8, 12 or 17 years old in the societies of today . . . To consult us would make your work more effective and give better results for the children. My proposal is that you make us part of your team. (Heidi Grande, 17, a Norwegian delegate to the Special Session on Children, SOWC, 2003, p. 65, my emphasis)

We are the children of the world, and despite our different backgrounds, we share a common reality. We are united by our struggle to make the world a better place for all. *You call us the future, but we are also the present.* (Final section of the statement presented to world leaders at the UN Special Session on Children, formulated between 400 young people at the Children's Forum, SWC, 2003, p. 66, my emphasis)

What is transparent from these three quotations direct from children and young people themselves is that they see themselves as *beings* in the here and now. They want action for them as they are now; they are keen to see change in their present times of 'being' rather than in their future times into which they are 'becoming'.

If the SOWC predominant discourse is that of children as 'becomings' what does this tell us about their perception of children and young people. It apparently indicates that the pervasive perspective of children as 'adults in waiting' is still very predominant even in UN discourses about children. UNICEF, in its own report about participation, in places represents such participation as *preparation* for future behaviour in adulthood. This 'slip' of conceptualisation is a classic example of the need for critical examination of the 'taken-for-granted' assumption within the text. If UNICEF, within its own report on participation, cannot effectively engage with the children as beings with rights in the here and now, then what hope is there of beleaguered and impoverished governments, communities and families doing the same? UNICEF has to set a meaningful example.

## Conclusions

There is a great deal of positive discourse in The State of the World's Children 2003 relating to the importance of authentic participation by children and young people. Examples are cited where individual children have made a significant intervention through their participation.

Nevertheless, there are facets of the document and its rhetoric which we have to critique. Such is the global reach of UNICEF's discourse that the vigilance of 'critical friends' is essential. Why are children and young people the holders of hopes and dreams? Is not this an indication that adults have to work harder at keeping in touch with, and striving for, their own hopes and dreams? Indeed, for many adults the hopes and dreams of a better future for their children and grandchildren is what keeps them striving against the odds. What about the children in such levels of despair that they have learned the futility of having hopes and dreams? How can these children begin the difficult path towards participation? Many of them are already participating in adult worlds (war, trafficking, prostitution) which have abused, exploited and damaged them. Where are the participation strategies for such children? Can 'authentic participation' reach children and young people who lack cultural capital, education, supportive homes and families? In the context of the specific report, what is not clear is whether the children feel their participation was taken seriously, whether they felt they were listened to at the 2002 meetings. Projects like the UNESCO 'Growing up in Cities' in conjunction with the MOST Programme and the work of Childwatch International (Auriat *et al.*, 2001) seem to indicate that it is possible. However, this is about reaching children and young people in cities. While the world is urbanising at a rapid rate (Boyden and Holden, 1991) there are millions of children in rural areas and in small island developing states. A more recent UNESCO programme, Small Island Voice, has a specific remit to promote and support children's and young people's participation through youth forums and other activities.

I remain concerned about the maintenance of a model of democracy which has already disenchanted and disenfranchised children and young people. Why cannot their encouraged participation be aimed at transformation rather than at confirmation of the established patterns? Is something of the vitality and creativity of children and young people lost when they participate in adult structures? If pre-existing models have marginalized children then unless there is fundamental change within the institutional structures children's participation will appear as tokenism, no matter how often this accusation is denied. Just as men were, and are, reluctant to give up their established forms of political (and other types) of power to allow women to play a meaningful role, so adults will resist the loss of authority and power that a child-centred, young person-friendly model of democracy will require. Change can be wrought and children and young people can grasp power (with adult support and advocacy) but it will be a long, long struggle. UNICEF, of all international organisations, should be the one to lead the way.

The young people and children involved in the Children's Forum and the UN Special Session raised their voices about a wide range of things. However, one of the most significant challenges was to encourage and demand that adults recognise children and young people as 'being' as opposed to always 'becoming'. A demand which is about recognition of the self and with that the confidence to demand rights and accept responsibilities in the here and now. It connects with new social studies of childhood and development discourses and practices of participation. However, it is important that UNICEF and other practitioners in development see children in their wider contexts. It is important that children are not seen as being so competent that we excuse ourselves as adults from the responsibility of caring for them. They *do* need protection and they *do* need provision. Just because some children participate and demonstrate social agency does not mean that they don't need looking after. We cannot ignore that the fact that the majority of the world's children/children of the majority world are in vulnerable positions which are *not* of their making. They are of adults' making, and frequently distant adults who fail or refuse to see the consequences of their actions.

Participation is an important part of children's rights. The CRC has created the space for a significant shift in thinking about, and working with, children. What is required now is an understanding from children's perspectives as to what they envisage as effective and meaningful participation and a thorough evaluation of what such participation might mean for those involved. Children's participation is firmly on the agenda but its practice has a long way to go.

There is evidence of such evaluations. A decade on from the launch of the CRC an international symposium on *Children's Participation in Community Settings'* was held in Oslo, Norway (June 2000). The emphasis was on best practice within programmes that provided children and young people with an authentic and effective voice as part of working towards an improvement in the life conditions, and this was enacted through the Growing Up in Cities project (Auriat *et al.*, 2001; Chawla, 2002a).

In my final section of participation in this debate I raise a very large question which I am not yet able to answer. It is a question triggered by three aspects; one is Boyden's (1990) criticism of the exclusion of voices from the South in the formulation of the CRC; a second is the critical analysis of the UNICEF report on children and participation outlined above; and the third reflects the fact the majority of the world's children live and die in so-called 'developing countries'. In consideration of these three aspects should the CRC be reviewed? Should an alternative Convention be written from the perspective of children, young people and their adult advocates of the majority world? If we are to genuinely challenge Northern/minority world hegemonic discourses around some of the most important people on the planet then my answer to both these questions is 'yes' and with a great deal of effort it could be a genuine and authentic process of participation.

## Acknowledgements

I was prompted to examine aspects of global discourses about children, young people and participation when I was invited to participate in an International Research Workshop entitled *The Global Child: Globalization, Development and Local Constructions of Childhood* funded by the Norwegian Centre for Child Research and the Department of Geography at the Norwegian University for Science and Technology held in March 2005. It was one of the most positive and stimulating academic experiences of my career. I wish to thank the convenors, Stuart Aitken, Anne Trine Kjørholt and Ragnhild Lund for their invitation, organisation and sumptuous hospitality and to the other participants for sharing their work, ideas and constructive comments. I am very grateful to anonymous referees of the paper for their important suggestions, but of course, I take full responsibility for the work itself.

## Notes

1. The ubiquity and pervasiveness of this definition became very apparent during a training session I organised as part of a Flinders University/AusAid gender mainstreaming programme in December 2005. The participants were all local government workers or NGO practitioners from different parts of Indonesia. I was presenting a final session at the end of a three-month programme which had focused on gender but not on children and young people in development. As part of the first stage of the session I asked them to define the term 'child'. I had expected different interpretations of the term, not least because of the variety of cultural and geographical backgrounds the participants represented. Also I was aware that few of them worked with children in their capacities as development workers. *All* 20 participants (in their groups of four or five) stated simply that a child was someone under 18.

2. The term 'the child' has been problematised here because it can objectify and universalise all children. Prout and James (1990) discuss the way in which Piaget used the term 'the child' as 'the bodily manifestation of cognitive development from infancy to adulthood [which] can represent all children' (pp. 12, my insertion).

3. Readers who are familiar with debates about gender and development will notice an echoing of the rhetoric defining the need for women's rights, autonomy, empowerment and participation within development.

4. However, geographers were not the initiators of this shift in conceptualisation of childhood within the social sciences. As Aitken states: 'Until very recently . . . geographers worked within commonly held assumptions about children, often without attending to the moral, cultural and political contexts of those assumptions' (2001, p. 27). Children's geographies is now, however, an important part of the 'new social studies of childhood'.

5. Once again the parallel debates with women's status within development are echoed here, in some quarters women are no longer seen as passive recipients of development but as social actors in their own right. However, a persistent view of their dependency on men remains strong.

6. This term is used in the State of the World's Children 2003 document and also defined in specific detail. It is something I return to later in the paper.

7. UNICEF's formalised discourse has progressively (but not completely) shifted from 'to children' to that of 'with children'.

8. There are those hopes and dreams again!

9. It is important to recognise that I am analysing just one Sate of the World's Children report. The 2005 report is titled 'Childhood Under Threat' which has a very strong emphasis on the impact of poverty on children. There are also repeated references to the importance of adults in children's lives.

10. This statement makes Jo Boyden's work on the 'globalization of childhood' all the more disturbing. Writing at a time pre-dating the CRC, Boyden argues that 'in recent consultations on the content of the draft of the Convention of the Rights of the Child . . . several delegates from the South expressed dissatisfaction that the drafting group was "predominantly Western in its orientation" and argued that greater account should have been taken "of the cultural diversity and economic realities of developing countries"' (1990, p. 198). The values of a so called 'normal' childhood were foreign concepts to them. With this insight we can see that the CRC is far from the democratic document the SOWC 2003 report presumes.

11. To be fair to UNICEF there is one specific mention of resistance in this 2003 report. It appears on page 17 and is the final 'reality' listing in Panel 2 *Child participation: myth and reality*. The report acknowledges that resistance can be an important part of participation and that it should be recognised as a form of communication and responded to through dialogue and negotiation not through force or persuasion.

12. Prout and James (1990, p. 11) have argued that within a Piagetian conceptualisation of children they are 'marginalized beings awaiting temporal passage, through the acquisition of cognitive skill, into the world of adults'. This is a problematic conceptualisation and connects with representations of children as passive, unable to articulate anything expect their own needs and desires.

13. This construction of children as 'becoming' resonates with youth studies work on transitions. For a more detailed critique of the 'youth transitions' concept and literature see Skelton (2002).

# References

Aitken, S. (2001) *Geographies of Young People*, London: Routledge.

Ansell, N. (2005) *Children, Youth and Development*, London: Routledge.

Auriat, N., Miljeteig, P. and Chawla, L. (2001) Overview: identifying best practices in children's participation, *Participatory Learning and Action*, pla notes, 42 downloaded from http://www.unesco.org/most/guic/guicpubframes.htm

Barrow, C. (2002) Introduction: Children's rights and the Caribbean experience, in C. Barrow (ed.) *Children's Rights, Caribbean Realities*, Kingston: Ian Randle Publishers, xiii–xxxiii.

Boyden, J. (1990) Childhood and the policy makers: a comparative perspective on the globalization of childhood, in A. James and A. Prout (eds) *Constructing and Reconstructing Childhood*, Basingstoke: The Falmer Press, 184–215.

Boyden, J. and Holden, P. (1991) *Children of the Cities*, London: Zed Books.

Cantwell, N. (1993) Monitoring the Convention through the idea of the '3 Ps', in P.L. Heiliö, E. Lauronen and M. Bardy (eds) *Politics of Childhood and Children at Risk: Provision, Protection and Participation*, Eurosocial Report 45, Vienna: European Centre for Social Welfare Policy and Research, 121–6.

Chawla, L. (ed.) (2002a) *Growing Up in an Urbanising World*, London/Paris: Earthscan Publications/UNESCO.

Chawla, L. (2002b) Cities for human development, in L. Chawla (ed.) *Growing Up in an Urbanising World*, London/Paris: Earthscan Publications/UNESCO, 15–34.

Dodman, D. (2004) Feelings of belonging? Young people's views of their surroundings in Kingston, Jamaica, *Children's Geographies*, 2(2), 185–98.

Foucault, M. (1980) *Power/Knowledge*, Brighton: Harvester.

Freeman, C., Nairn, K. and Sligo, J. (2003) 'Professionalising' participation: from rhetoric to practice, *Children's Geographies*, 1(1), 53–70.

Freeman, M.D. (1992) The limits of children's rights, in M.D. Freeman and P. Veerman (eds) *The Ideologies of Children's Rights*, International Studies in Human Rights, Dordrecht: Nijhoff Publishers, 29–46.

Hammarberg, T. (1990) The UN Convention, the rights of the child and how to make it work, *Human Rights Quarterly*, 12, 97.

Hart, R.A. (1992) *Children's Participation: From Tokenism to Citizenship*, Innocenti essay, No 4, UNICEF International Child Development Centre, Florence, Italy.

Hart, R.A. (1997) *Children's Participation, the Theory and Practice of Involving Young Citizens in Community and Environmental Care*, New York/London: UNICEF/Earthscan.

Holloway, S. and Valentine, G. (2000) Children's geographies and the new social studies of childhood, in S. Holloway and G. Valentine (eds) *Children's Geographies: Playing, Living, Learning*, London: Routledge.

James, A. and Prout, A. (eds) (1990) *Constructing and Reconstructing Childhood*, London: Falmer Press.

James, A., Jenks, C. and Prout, A. (1998) *Theorising Childhood*, Cambridge: Polity Press.

Johnson, V., Ivan-Smith, E., Gordon, G., Pridmore, P. and Scott, P. (eds) (1998) *Stepping Forward: Children and Young People's Participation in the Development Process*, London: Intermediate Technology Publications.

Katz, C. (2004) *Growing up Global*, Minnesota, MN: University of Minnesota Press.

Kelley, N. (2006) Children's involvement in policy formation, *Children's Geographies*, 4(1), 37–44.

Kjørholt, A.T. (2001) The participating child—A vital pillar in this century? *Nordisk Pedagogikk/Nordic Educational Research*, 21(2), 65–81.

Kjørholt, A.T. (2002) Small is powerful: Discourses on 'children and participation' and Norway, in *Childhood*, 9(1), 63–82.

Kjørholt, A.T. (2004) *Childhood as a Social and Symbolic Space: Discourses on Children as Social Participants in Society*, published doctoral thesis, Norwegian Centre for Child Research, Trondheim, NTNU.

Matthews, H. (2005) The millennium challenge: The disappointing geographies of children's rights, *Children's Geographies*, 3(1), 1–3.

Matthews, H., Limb, M. and Taylor, M. (1999) Young people's participation and representation in society, *Geoforum*, 30, 135–44.

Mayo, M. (2001) Children's and young people's participation in development in the South and in urban regeneration in the North, *Progress in Development Studies*, 1(4), 279–93.

McNeish, D. and Gill, T. (2006) Editorial: UK policy on children: key themes and implications, *Children's Geographies*, 4(1), 1–7.

Nelson, N. and Wright, S. (1995) Participation and power, in N. Nelson and S. Wright (eds) *Power and Participatory Development*, London: Intermediate Technology Publications, 1–18.

Percy-Smith, B. (2002) Contested worlds: constraints and opportunities in city and suburban environments in an English Midlands city, in L. Chawla (ed.) *Growing Up in an Urbanising World*, London/Paris: Earthscan Publications/UNESCO, 57–80.

Prout, A. and James. A. (1990) A new paradigm for the sociology of childhood? Provenance, promise and problems, in A. James and A. Prout (eds) *Constructing and Reconstructing Childhood*, Basingstoke: The Falmer Press, 7–34.

Qvortrup, J. (1994) *Childhood Matters: Social Theory, Practice and Politics*, Aldershot: Avebury.

Rose, G. (2001) *Visual Methodologies: An Introduction to the Interpretation of Visual Materials*, London: Sage Publications.

Sibley, D. (1995) *Geographies of Exclusion*, London: Routledge.

Skelton, T. (1999) Representations of Jamaican yardies: domination and resistance, in J. Sharp, C. Philo, P. Routledge and R. Paddison (eds) *Entanglements of Power: Geographies of Domination and Resistance*, London: Routledge, 182–203.

Skelton, T. (2000) 'Nothing to do, nowhere to go?': Teenage girls and public space in the Rhondda Valleys, South Wales, in S.L. Holloway and G. Valentine (eds) *Children's Geographies: Playing, Living, Learning*, London: Routledge, 80–99.

Skelton, T. (2002) Research on youth transitions: some critical interventions, in M. Cieslik and G. Pollock (eds) *Young People in Risk Society: The Restructuring of Youth Identities and Transitions in Late Modernity*, Ashgate: Aldershot, 100–16.

Stephens, S. (1995) Children and the politics of culture in late capitalism, in S. Stephens (ed.) *Children and the Politics of Culture*, Princeton, NJ: Princeton University Press, 3–48.

UN (1989) *Convention on the Rights of the Child*, Secretary-General, New York: UN.

UNESCO (2002) *UNESCO-Mainstreaming the Needs of Youth*, Paris: UNESCO.

UNESCO (2004) *Small Islands Voice: Laying the Foundations*, Paris: UNESCO.

UNICEF (1990) *Convention on the Rights of the Child, First Call for Children*, New York: UNICEF, 41–75.

UNICEF (2002) *The State of the World's Children* available as PDF at the following web site: http://www.unicef.org/publications/files/pub_sowc03_en.pdf

Waitt, G. (2005) Doing discourse analysis, in I. Hay (ed.) *Qualitative Research Methods in Human Geography*, Melbourne: Oxford University Press, 2nd edition.

Weller, S. (2006) Situating (young) teenagers in geographies of children and youth, *Children's Geographies*, 4(1), 97–108.

# Index